Diffraction Grating Spectrographs

Diffraction Grating Spectrographs

Sumner P. Davis

University of California, Berkeley

HOLT, RINEHART AND WINSTON
NEW YORK   CHICAGO   SAN FRANCISCO   ALTLANTA   DALLAS
MONTREAL   TORONTO   LONDON   SYDNEY

Preface

An experimentalist is often faced with the problem of choosing a spectrograph to use in his research. The basis for a proper choice is sometimes unknown to him, since his experience may not include a broad acquaintance with optical equipment. This book seeks to show the scientist how to make the best choice of a diffraction grating instrument for his particular research problem. It is not intended to be a complete exposition of theory and practice, but many practical points of operation are touched upon, which are not explained in other texts. Anyone who makes a choice and then sets out to use a spectrograph must of necessity refer to more detailed expositions in order to learn the appropriate experimental technique. Adequate though far from complete references are given for this purpose. Techniques vary, of course, and no two people work in the same way. This exposition is based on my own background and experience, and strongly reflects my association with other spectroscopists. The reader is expected to be thoroughly familiar with optics and spectroscopy at the level of G. R. Fowles, *Introduction to Modern Optics* (Holt, Rinehart and Winston, New York), or F. A. Jenkins and H. E. White, *Fundamentals of Optics* (McGraw Hill, New York), to which he may refer for more detailed explanations and illustrations of some of the definitions used here.   George R. Harrison and Joseph Reader have been most helpful with their critical comments.

S. P. D.

Berkeley, California
November, 1969

Contents

Diffraction Grating Spectrographs

Fundamental Properties and Equations

INTRODUCTION

The optical spectrum of electromagnetic radiation is arbitrarily separated into several regions as shown in Table 1. It is well to remember both the wavelength and wavenumber limits of each region, and to think in both terms. Many units are now in vogue, and it seems unlikely that any one set will win out over the others. Wavelengths are specified in meters (m), centimeters (cm), millimeters (mm), micrometers, microns (μ, 10^{-6} m), millimicrons (mμ, 10^{-9} m), nanometers (nm, 10^{-9} m), and angstroms (Å, 10^{-10} m). Wavenumbers—the reciprocals of vacuum wavelengths— are given in reciprocal meters (m^{-1}), reciprocal centimeters (cm^{-1}), and kaysers (K, cm^{-1}).

The location of a spectral line is given by its wavelength in air (except in the vacuum ultraviolet region), or by its wavenumber in vacuum. When small wavelength or wavenumber differences are discussed, the intervals are given in millikaysers (mK, 10^{-3} cm^{-1}) or in megahertz (MHz, 10^{9} sec^{-1}; 30 MHz is equivalent to 1 mK).

TABLE 1
Regions of the Optical Spectrum

	Wavelength Range	Wavenumber Range
Vacuum ultraviolet	5 Å– 2,000 Å	2×10^7 cm^{-1}–50,000 cm^{-1}
Ultraviolet	2,000 Å– 4,000 Å	50,000 cm^{-1}–25,000 cm^{-1}
Visible	4,000 Å– 7,000 Å	25,000 cm^{-1}–14,300 cm^{-1}
Photographic infrared	7,000 Å–12,000 Å (1.2μ)	14,300 cm^{-1}– 8,333 cm^{-1}
Infrared	1.2μ–40μ	8,333 cm^{-1}– 250 cm^{-1}
Far infrared	40μ–1000μ	250 cm^{-1}– 10 cm^{-1}

The two fundamental uses of a spectrograph are to separate and measure wavelengths present in electromagnetic radiation, and to measure the relative amounts of radiation at each wavelength.

There are three types of instruments, each based on a different physical phenomenon.   The prism spectrograph uses the dispersion of a transparent medium to separate wavelengths.   The diffraction grating utilizes wavefront division and interference of the radiation for its dispersive properties, while the interferometer depends upon amplitude division of the wavefront.

All three types of spectrograph can be described by several characteristic properties.   The *resolving power* is a number characterizing the property of separating, or resolving, or recording as distinct two monochromatic input radiations of the same intensity and of nearly the same wavelength.   An arbitrary criterion is established for this separation. When this criterion is satisfied, the wavelength difference is denoted by $\delta\lambda$, the wavenumber differences by $\delta\sigma$, and the resolving power $R = \dfrac{\lambda}{\delta\lambda} = \dfrac{\sigma}{\delta\sigma}$. Sometimes it is more significant physically to use the *resolving limit* itself, $\delta\sigma$, rather than R, as a measure of the resolution of the spectrograph.

Every spectrograph deviates incident light through an angle θ which depends upon the wavelength λ.   The derivative $\dfrac{d\theta}{d\lambda}$ is called the *dispersion*. However, it is more common to use the reciprocal dispersion or *plate factor* $\dfrac{d\lambda}{dl}$ as a characteristic function of the instrument.   The quantity l is the distance from an arbitrary reference mark at the output, to the position where radiation of wavelength λ appears.

The *free spectral range* is the interval of the spectrum that can be observed without interference or overlapping from other wavelengths in the incident radiation.   The definition and importance of this quantity will become clearer as the individual instruments are discussed.   It is denoted by F_σ when expressed in wavenumbers and F_λ when expressed in wavelengths.

The *throughput* of a spectrometer is a measure of the detector response for a given light source. There is no specific number for this response; the measurement is a relative one, comparing the responses of two instruments when the same light source and detector are used. The throughput depends first upon the speed of the instrument—how large a solid angle of light it will accept, which is given approximately by the square of the aperture ratio. Second, it depends upon the transmission factor—what fraction of the light entering the instrument reaches the detector. Third, it depends on the sizes of the entrance and exit apertures.

SPECTROGRAPH OPTICS

A spectrograph contains two kinds of optics. The first kind forms images of the source. The second disperses the light, or makes the angular deflection of the light a function of wavelength. Together these make an instrument that forms many images of the source, one for each wavelength present in the radiation. A general schematic set-up is shown in Fig. 1. The source is chosen to be a narrow entrance slit.

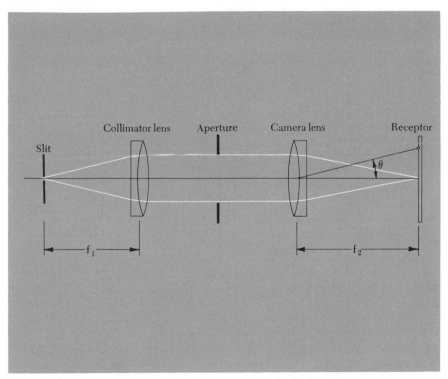

FIG. 1. Image forming optics of prism and grating spectrographs.

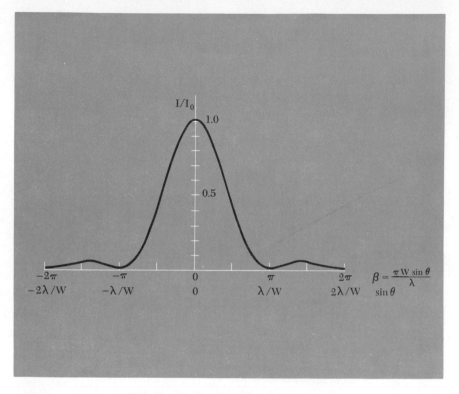

FIG. 2. Single slit diffraction pattern.

Let us assume that the slit is of infinitesimal width.   Then its image at the receptor in the focal plane of the camera lens is a diffraction pattern, whose width is dependent upon the size W of the aperture of the optical system.   The distribution of monochromatic light of wavelength λ at the angle θ is given by

$$I(\theta) = \text{constant} \times \left| \int_{\substack{\text{aperture} \\ \text{width}}} e^{iky \sin \theta} \, dy \right|^2 = I_0 \frac{\sin^2 \beta}{\beta^2}$$

$I(\theta)$ = power per unit area per unit wavelength interval = intensity

$$\beta = \frac{\pi W \sin \theta}{\lambda}$$

This function is the well-known single-slit diffraction pattern with a width characteristic of the aperture width, and is shown in Fig. 2.   The larger the aperture the narrower the central image, and vice-versa.

If a dispersing element is placed at the aperture, single slit diffraction patterns are formed at various angles θ (positions at the receptor), with each image corresponding to a particular wavelength or color of light.

DIFFRACTION GRATINGS

When a diffraction grating is placed in the aperture of the instrument schematically illustrated in Fig. 1, a grating spectrograph is the result. Slit images are formed at angles corresponding to wavelengths present in the incoming radiation. Let us choose for illustration a plane transmission grating, as shown in Fig. 3. The same basic equation applies:

$$I(\theta) = \text{constant} \times \left| \int e^{iky \sin \theta} \, dy \right|^2$$

$$= \text{constant} \times \left| \int_{-a/2-a_0/2}^{-a/2+a_0/2} e^{iky \sin \theta} \, dy + \int_{a/2-a_0/2}^{a/2+a_0/2} e^{iky \sin \theta} \, dy + \cdots \right|^2$$

$$I(\theta) = \text{constant} \times \left| \int_{-a_0/2}^{a_0/2} e^{iky \sin \theta} \, dy \{ 1 + e^{ika \sin \theta} \right.$$

$$\left. + e^{i \, 2ka \sin \theta} + \cdots + e^{i(N-1)ka \sin \theta} \} \right|^2$$

$$I(\theta) = \text{constant} \times (Na_0)^2 \left(\frac{\sin \beta}{\beta} \right)^2 \left(\frac{\sin N\gamma}{N \sin \gamma} \right)^2$$

$$I(\theta) = \text{constant} \times W^2 \left(\frac{a_0}{a} \right)^2 \left(\frac{\sin \beta}{\beta} \right)^2 \left(\frac{\sin N\gamma}{N \sin \gamma} \right)^2$$

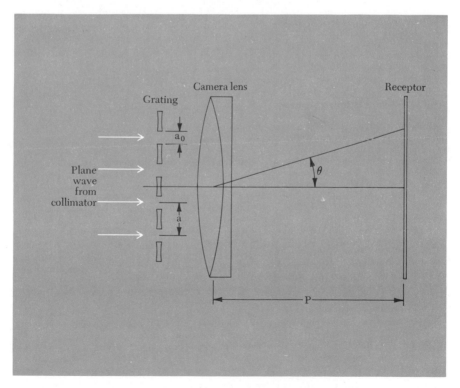

FIG. 3. Diffraction at a transmission grating.

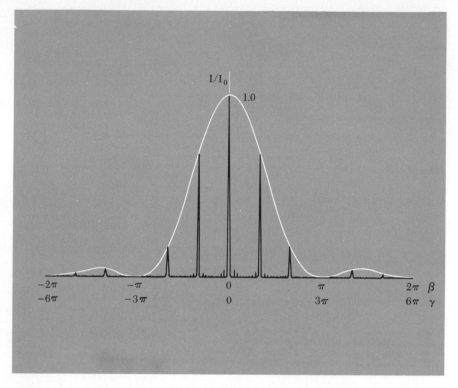

FIG. 4. Diffraction pattern of a transmission grating with 10 slits and $a = 3a_0$.

N = total number of apertures

$\beta = (\pi a_0 \sin \theta)/\lambda$

$\gamma = (\pi a \sin \theta)/\lambda$

W = width of grating

a_0 = width of apertures

a = center-to-center spacing of apertures

This latter equation is the interference pattern of a diffraction grating, as shown in Fig. 4.   Let us look at the equations more carefully.

The maximum intensity is proportional to W^2.   One factor of W arises because a larger grating accepts a larger solid angle of light.   The second factor arises because the image width gets smaller as the aperture width gets larger.   The factor $\sin^2 \beta/\beta^2$ tells us that there is a single-slit diffraction pattern characterized by the width a_0 of a single aperture; the factor $\left(\dfrac{\sin N\gamma}{N \sin \gamma}\right)$ is the N-slit interference pattern, from which is derived the grating equation $m\lambda = a \sin \theta$ (for light striking the grating at normal incidence and diffracted at an angle θ).   This equation gives the locations of the principal maxima of the function; these maxima are known as orders of the spectrum.   The order m takes on integral values, positive or negative.   When $m = 1$, then there is just one wavelength path difference between light which reaches the point P in the receptor plane from adjacent slits, as indicated in Fig. 5.   There are secondary maxima which can theoretically have up to 4% of the intensity of the principal maxima.   In practice they are never seen, for one or more of the following reasons: the actual line shapes are considerably broader than theory indicates, the sources are not strictly monochromatic, the dispersion is not large enough, or the slit width is not small enough.

RESOLVING POWER

Two spectral lines of nearly equal wavelengths and of equal intensities are said to be resolved according to the Rayleigh criterion when the maximum intensity of one interference pattern falls on the zero minimum of the other interference pattern, as shown in Fig. 6.   This condition occurs when $\left(\dfrac{\sin N\gamma_1}{N \sin \gamma_1}\right)$ is a maximum and $\left(\dfrac{\sin N\gamma_2}{N \sin \gamma_2}\right)$ is a minimum at the same angle θ,

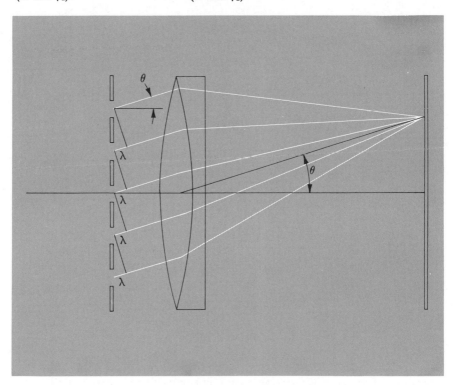

FIG. 5.   Constructive interference producing the first order spectrum.

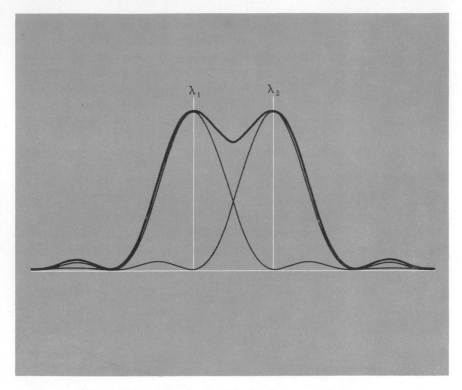

FIG. 6. Rayleigh criterion for chromatic resolving power.

or

$$N\gamma_1 = mN\pi = \frac{N\pi a \sin\theta}{\lambda_1} \text{ and } N\gamma_2 = (mN + 1)\pi = \frac{N\pi a \sin\theta}{\lambda_2}$$

or

$$mN\lambda_1 = (mN + 1)\lambda_2$$
$$R = \frac{\lambda}{\delta\lambda} = \frac{\sigma}{\delta\sigma} = mN$$

where m is the order of interference and N the number of apertures.   The equation can be transformed as follows:

$$R = mN = \left(\frac{a \sin\theta}{\lambda}\right)\left(\frac{W}{a}\right) = \frac{W \sin\theta}{\lambda}.$$

This number is numerically equal to the total number of wavelengths in the path difference between the extremes of the grating.   Note that for a given angle and wavelength, the resolving power is independent of groove spacing.   To increase the resolving power of a grating, it is evident that the grating must be made larger, or the angle of diffraction must be made larger, or both.   The resolving limit $\delta\sigma$ is just $1/(W \sin \theta)$.

DISPERSION

The dispersion is given by $\dfrac{d\theta}{d\lambda} = \dfrac{\tan \theta}{\lambda}$ and the plate factor $\dfrac{d\lambda}{dl} = \dfrac{d\lambda}{Pd\theta} = \dfrac{\lambda}{P \tan \theta}$.   The focal distance P is used instead of the focal length f, since not all grating mountings have a fixed focal distance equal to f.

FREE SPECTRAL RANGE

The free spectral range can be determined by setting $m\lambda_1 = (m + 1)\lambda_2$. When the difference in wavelengths is small compared to λ_2,

$$\lambda_1 - \lambda_2 = F_\lambda \simeq \frac{\lambda}{m} = \frac{\lambda^2}{a \sin \theta}.$$

The corresponding quantity in wavenumbers is

$$F_\sigma = \frac{\sigma}{m} = \frac{1}{a \sin \theta}.$$

THROUGHPUT

The f/number of the optics of a grating spectrograph is given by the projected width of the grating divided by the focal distance.   Small instruments, for example, are often made as fast as $f/6$.   However, in the case of the transmission grating illustrated in Fig. 3 only a small part of the light that is incident on the grating is dispersed into a useful spectrum.   Most of the light goes straight through into the zero order, and is wasted so far as the formation of the spectrum is concerned.   On the other hand, it is possible to *blaze* gratings, and to throw most of the light into a single order of a specified wavelength.   To accomplish this in the case of a transmission grating, a small linear phase shift is introduced at each slit, zero at the top and ϕ_0 at the bottom, for example, as shown in Fig. 7(a).   As light of a particular wavelength λ_0 passes through each slit, it is refracted through an angle θ_0.   But, the light from each successive slit has traveled some distance farther than has that from the preceding one.   If this distance is just one wavelength, then at the angle θ_0 there appears the first order interference maximum.   The grating is said to be blazed at the first order for the wavelength λ_0 at the angle θ_0; the central maximum of the single slit diffraction pattern and the first maximum of the N-slit interference pattern coincide.   Hence, most of the energy is in the first order spectrum.   If a wavelength other than λ_0 is incident on the grating, then there is no longer a single order with appreciable light, but two orders. Both these cases are illustrated in Fig. 7(b).

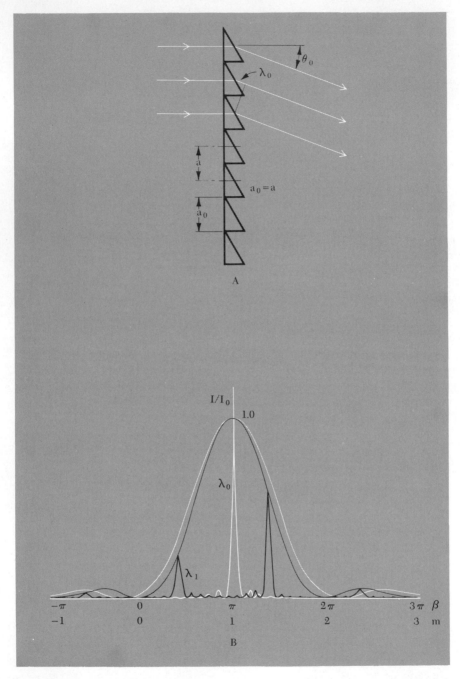

FIG. 7. (a) Blazed transmission grating. (b) Diffraction pattern of a blazed transmission grating.

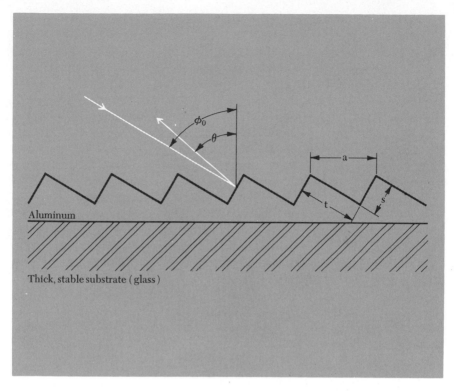

FIG. 8. Profile of a reflection grating.

Almost no transmission gratings are used in research instruments, even though it has been claimed that practically obtainable blazes throw up to 90% of the light into a single order for a specified wavelength. Some of the reasons are that the dispersion of the grating material introduces further complexities, materials begin to absorb in the ultraviolet region of the spectrum, and it is difficult and expensive to make image forming transmission optics which are achromatic over a wide wavelength range. Also, techniques for making well blazed transmission gratings were not developed until recent years. All research quality gratings are of the reflection type, where light is reflected from a mirror surface ruled with grooves, rather than transmitted through slits. The principle of operation is the same as already described, except that the light is reflected as well as diffracted at the grating surface. The grooves are specially shaped so as to reflect the light into the desired order of interference. When the grooves are sharply defined as shown in Fig. 8, the grating is said to be strongly blazed. They most often have one steep and one shallow face, with an angle between the faces of about 90 degrees. The light is arranged so as to strike the grating at an angle of incidence ϕ_0, parallel to the shallow face of width t and perpendicular to the steep face of width s. The diffracted light is observed at an angle ϕ nearly equally to the angle of incidence. This condition of use is known as autocollimation. The grating equation is $m\lambda = a(\sin \theta + \sin \phi_0)$ and of course includes both the angle of incidence and the angle of diffraction. The single slit diffraction pattern is centered about the angle $\theta_0 = \phi_0$ from the normal to the grating, and has an angular width dependent upon the distance s. The interference maxima have positions and spacings dependent on the distance $a(\sin \theta + \sin \phi_0) \simeq 2t$. If $2t$ is an integral number of wavelengths, then the pattern appears similar to that shown in Fig. 7(b), and most of the light is thrown into the single order $m = 2t/\lambda$. The mth order interference maximum falls at the central maximum of the single slit diffraction pattern, while all other orders fall at zero minima of the diffraction pattern. At a different wavelength, only two orders have much intensity, and all the rest are weak, as is also illustrated in Fig. 7(b). A grating is blazed over a wavelength range $\Delta\lambda = \pm\lambda/2m$, according to the following considerations. Referring to Fig. 8, let θ and φ be the same and equal to θ_0, and $m\lambda = 2a \sin \theta_0$. Light of appreciable intensity is diffracted only up to about one-half an angle δ each side of θ_0, given by the width of the central diffraction maximum of a single slit, $\sin \delta = \pm\lambda/s$. The wavelength at the edge of the diffraction pattern is given by

$$m\lambda' = a[\sin (\theta_0 + \delta) + \sin \theta_0]$$

$$m\lambda' \simeq 2a \sin \theta_0 + a \cos \theta_0 \sin \delta$$

$$m\lambda' \simeq m\lambda + a \left(\frac{s}{a}\right)\left(\frac{\lambda}{s}\right)$$

$$\lambda' - \lambda \simeq \lambda/m.$$

Taking one-half this value,

$$\Delta\lambda \simeq \pm\lambda/2m.$$

Practically Obtainable Diffraction Gratings

INTRODUCTION

So far we have directed our attention to the fundamental properties of gratings from a theoretical standpoint, but have said little about practically attainable gratings, and how close they come to exhibiting their theoretical properties.[1] A diffraction grating is usually ruled by a diamond in a thin coating of aluminum which has been deposited on a glass blank. A glass (or fused silica) blank of proper dimensions to be stable is polished to a flatness of a fraction of a wavelength. Then a coating of aluminum is vacuum evaporated on the surface to a thickness varying from about 0.4 microns for a groove spacing of 1/1200 mm, to about 18 microns for a

[1] The art of ruling gratings has been practiced for many years, but the most successful developer of the science has been George R. Harrison, of The Massachusetts Institute of Technology. His introduction of interferometric control of the ruling process has been the key to success. For a number of years he and his associates have consistently ruled the largest and most accurately spaced gratings.

spacing of 1/25 mm.   When coating thicknesses are much larger than one micron, many irregularities occur, and the surface often shows evidence of crystallization and exhibits cosmetic (surface) blemishes.   The grating is ruled by a boat-shaped diamond, which pushes aside the aluminum as if it were butter.   The time required for ruling can be quite long.   For example, an engine may rule 12 grooves/minute, or 2.4 mm/hr at a spacing of 1/300 mm.

The size of a grating depends upon the length of time and the distance over which the accuracy of the ruling can be controlled.   The length of groove is limited only by the mechanical construction of the engine. Gratings ruled on wholly mechanical engines rarely have greater than 15 cm of usable width; those ruled on interferometrically controlled engines have been made with excellent quality up to 25 cm, and the prospects are now good for much larger ones.   Most of the development of better ruling methods has taken place with plane gratings, and since about 1948 attention has been paid to improving plane grating spectrographs, until at the present time (1969) most of the new spectrographs use plane gratings, except in the vacuum ultraviolet where the use of concave gratings predominates.

GROOVE SPACINGS

The limitation on groove spacing is not the shape of the diamond, but the length of time which the engine will operate stably (limits maximum number of grooves) and the thickness of the aluminum coating (limits the minimum number of grooves/mm.   Large thicknesses tend to become grainy and crystalline, and hence poorly reflecting).   Of course, it makes no sense to rule grooves more closely spaced than the wavelength of light to be used.   The upper practical limit is 4000/mm; the lower limit is about 25/mm at the present time.

RULING ERRORS AND THEIR CONSEQUENCES

A periodic sinusoidal error in groove spacing is the most common characteristic of gratings produced on strictly mechanical engines.   The carriage holding the grating blank is advanced by a screw; any periodic error in the screw threads or eccentricity in mounting the screw results in a periodic error, which causes false spectrum lines spaced regularly on each side of a true spectrum line.   An illustration is shown in Fig. 9(a).   These false lines are called Rowland ghosts; their apparent wavelengths are given by $\lambda' = \lambda(1 \pm m'/mn)$ where λ is the wavelength of the spectral line, m is the order, m' the order of the ghost, and n the number of grooves ruled in one turn of the screw.   Good gratings have ghost intensities less than $1 \times 10^{-3}\%$ of the spectrum line intensity, in the first order.   The

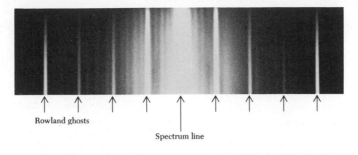

Rowland ghosts

Spectrum line

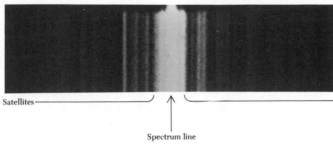

Satellites

Spectrum line

FIG. 9. Spectrograms showing: (a) Rowland ghosts and (b) Satellites of a spectrum line.

ghost intensity increases as the square of the sine of the angle of diffraction, and the inverse square of the wavelength. The best gratings ruled on interferometrically controlled engines show no measurable Rowland ghosts.

While the engine rules, there may be random displacements of the diamond relative to an absolute standard of groove spacing. Sometimes these displacements last for a few grooves, but occasionally for several millimeters of ruling. Each of these sections acts as a small grating, and each produces a broad spectral line somewhat displaced from the true line. The net result is to produce a broadened spectrum line when the errors are small, and many relatively weak satellite lines when the errors are large. Many an otherwise good grating has had to be discarded because of the presence of these satellite lines. They are, of course, particularly troublesome when hyperfine structure is being observed, or any spectrum which has both very intense and very weak lines. Under these conditions it is difficult to separate false lines from true ones. An illustration is shown in Fig. 9(b).

It should be noted that the presence of ghosts and satellites need not seriously degrade the resolution of the grating. In fact, some gratings showing excellent resolution for weak lines have such strong well-resolved satellites that they are not very useful in practice.

Lyman ghosts are false lines well separated from the true spectrum

line, often by a large fraction of an order, and can have several percent of
the intensity of the parent line in high orders of interference.   Their cause
is not yet thoroughly explained, but appears to be connected with vibrations of the diamond during ruling.

In the early days of ruling, there were several other ruling errors which
caused false spectrum lines to appear.   Now, they are of little more than
historical interest, and hence will not be mentioned here.

BLAZE

The blaze of a grating or its efficiency is highly important, since no grating
is useful unless it throws a measurable quantity of light into a useful order.
The efficiency is a measure of how much of the incident light of a particular
wavelength is thrown into a single order.   The amount depends not just
upon the quality of the reflecting surface of the grating, but on the shape
of the groove faces.   Ideally, when used at the blaze angle, the grating
looks just like a plane mirror.   But, in ruling, the diamond cannot cut a
perfect groove; in fact it pushes aside the aluminum, rather than cuts it
out, and the excess aluminum flows over the sharp edge of the previous
groove.   One face of the groove has a rounded part instead of being wholly
plane, and consequently throws light into other angles than the blaze.   A
well-ruled grating has two blaze angles, one on each side of the normal, but
of unequal efficiencies.   The weaker blaze occurs when the faces are used
which the aluminum flowed over the edge during ruling.

The grating is most efficient when used exactly at the blaze angle,
since the entire face is utilized.   When used at a different angle, the sides
of the grooves interfere with either the incident or diffracted light, and hence
diminish the useful area.   Also, light which strikes the sides of the grooves
is reflected into other regions of the spectrum.   Gratings with small groove
spacings must be used over wide ranges of incidence and diffraction in order
to observe the entire optical spectrum, while gratings with large spacings
can be used over a correspondingly smaller range of angles, and hence are
more efficient throughout the range.   The echelle grating spectrograph is
designed to take advantage of these blaze properties.

Taking these facts into consideration, it can be said that a grating
with a smaller spacing has a less pronounced blaze, and disperses the
spectrum with more nearly equal intensities at all wavelengths.   A grating
with a larger groove spacing has a more pronounced blaze, and is useful
over only a restricted range of angles.   Because of the larger spacing and
consequent smaller free spectral range, however, a smaller range of angles
is needed to observe the spectrum.   Further mention is made of these
properties in the discussions of concave and plane gratings.

TESTING OF GRATINGS

A research grating should be given a thorough testing before being used, so that the experimenter knows what it will do, and has confidence in it. There are several tests which can be applied, all of which are important, but unfortunately no single manufacturer employs all of them. It is often up to the purchaser to arrange for the requisite tests, or to perform them himself.

The most important test is of the planarity of the wavefront diffracted from the grating. If it is a perfectly plane wave, the quality of the image will be determined by the image forming optics of the spectrograph. If it is not, various aberrations and errors in the spectrum will result. To make this test, the grating is set up in place of one reflecting mirror in a Michelson or a Fizeau interferometer. The wavefront returned from the grating is tilted (by adjustment of the grating) to a slight angle with the wavefront from the plane reference mirror. The intersection of these two wavefronts produces fringes, which can be observed visually and photographically. If both wavefronts are plane, the intersections are straight lines, and the grating is perfect. If the reference mirror is assumed to be flat to the tolerances required, then any deviation from straightness is a direct measure of ruling errors. The exact effect of these errors on the final image must be calculated—which the experimenter is not prepared to do, but the qualitative effect can easily be deduced. A general curvature indicates a curved blank (a common difficulty with large groove spacings, where the aluminum coating is thick), and astigmatism in the final image will result (but no degradation of resolution if the curvature is spherical). Sudden breaks in the fringes indicate sudden changes in groove spacing. If the fringes return to their normal path, then there is a section with a slightly different spacing, and a satellite is the result; if the fringes do not return to their paths, a second grating is the result. A glance at the fringe pattern will show whether the grating is worth testing further or not. Examples are shown in Fig. 10.

For the next tests, the grating should be set up in a spectrograph; a convenient mounting is either a Littrow or a Czerny–Turner mounting. It is important to remember that the plate factor of the spectrograph must be sufficiently large to make full use of the grating's properties. For a 12 cm wide grating, this requirement means a focal length of about 6 m, and for a 25 cm grating, a focal length of 12 m. Reasons for this are explained later. A low pressure mercury arc lamp should be used as a source and the 5461 Å line observed. With the grating set at the blaze, and the spectrograph in focus, place your eye at the slit image on the plateholder and look at the face of the grating. The parts of the grating which are illuminated show just what sections are contributing to the spectral image.

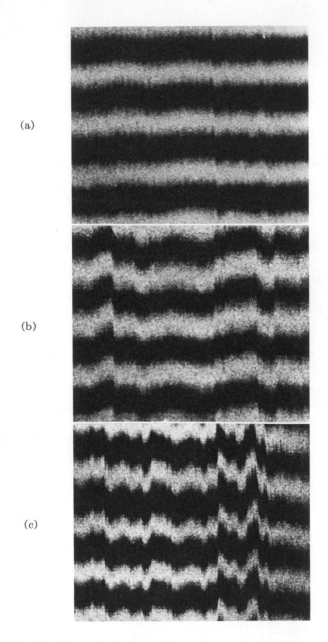

(a)

(b)

(c)

FIG. 10. Fringes form from the wavefront of a diffraction grating. (a) excellent, (b) good, (c) fair grating. (Interferograms courtesy of George R. Harrison.)

A good grating is uniformly illuminated and is said to show no target pattern.   See Fig. 11 for an example.   However, even if the grating is not uniformly illuminated, it doesn't mean that it is unusable; it just means that the groove shape is not uniform over the entire surface of the grating, and hence that the wavefront is not of uniform intensity over the entire surface.   The resolution will be somewhat degraded, and the blaze weaker than it should be.   Remember that a narrow slit partially polarizes the light, so that this test should be carried out with a slit which is at least several wavelengths wide.   The grating also polarizes the light, particularly if the groove spacing is small, and the grating is used at a very high angle.

The blaze can be tested in the same mounting.   The slit is set for an appropriate value to give enough light, and a photomultiplier without an

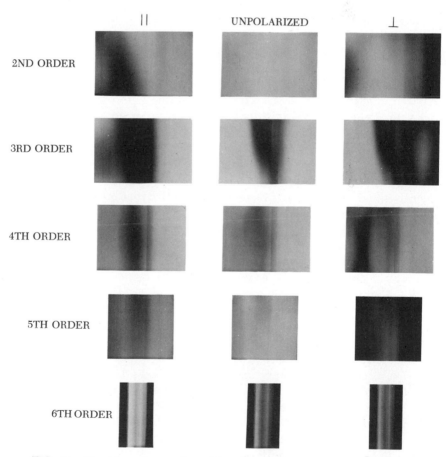

FIG. 11.   Target patterns of a diffraction grating.   [After A. Keith Pierce, *J. Opt. Soc. Am.* **47**, 6 (1957).]

exit slit is used as the receptor. The response of the photomultiplier can be compared with the response from a plane mirror of the same aperture as the grating. When the grating is rotated to bring successive orders into view, the slit width must be changed in order to keep the same total amount of light incident on the grating. It will be found that gratings throw light into more than one order when tested as above, with one usually much stronger than the rest. The grating will be blazed on both sides, as explained earlier. It will also be noted that orders lower then the blaze (smaller angles) are reasonably bright, while orders above the blaze drop sharply to nearly zero intensity, and hence are not useful. The blaze should be measured for wavelengths over the entire range expected to be used, since the angles for maximum light may be somewhat different for different wavelengths.

The test for ghosts and satellites is easily made if the spectrum can be scanned continuously over a small wavelength range. The grating is set up as above, but this time the photomultiplier is fitted with a slit adjusted to give the theoretical resolution of the grating (together with proper adjustment of the entrance slit). A spectrum line is then scanned, and the output of the photomultiplier put on a strip chart recorder. The relative intensities can then be directly measured. In this test a sharp line source must be used. An ordinary mercury arc may not give sharp enough lines both because of the temperature at which it operates and the presence of several isotopes. A better source is a mercury 198 electrodeless lamp; even here care must be exercised, since there is always a small amount of other isotope contamination, and small apparent satellites appear with a good grating which are legitimate hyperfine structure. If a means of scanning is not available, photographic exposures can be taken in the usual way with differing times, in order to calibrate the plate and make relative intensity measurements possible.

The measurement of scattered light is difficult to make quantitatively, and is often not necessary. Qualitative tests can be made, but the usually satisfactory practice is to set up the grating in its proposed operating condition, and try it out. The problem is greatest when weak lines are to be measured at the same time as strong ones, or when there is considerable light in some portion of the spectrum, as may be the case when observing a band spectrum with intense heads.

The testing of concave gratings generally follows the same procedure as for plane gratings, with appropriate modifications for the different conditions of use. An auxiliary spherical mirror is needed when testing for wavefront aberrations, in order to render the diffracted wavefront plane. The target pattern, blaze, ghost, and satellites are measured using methods similar to those for plane gratings.

TABLE 2
*Bausch and Lomb Grating Selection Chart**

Approximate Spectral Band	Sizes and Types of Gratings	Grooves/mm	Blaze	Remarks
Soft x-ray 10 Å to 250 Å	Concave reflection gratings used in high vacuum at grazing incidence; sizes and types with radii from 400 to 6650 mm.	600–3600/mm	1° to 6°	Al, Au, Pt coatings. Grazing Incidence
Extreme Ultraviolet (XUV) 250 Å to 1100 Å	Concave reflection gratings used near normal incidence; usually in first order.	600–3600/mm	1° to 10°	Au replicas or Pt flash on Al.
Vacuum Ultraviolet 1100 Å to 2000 Å	Large variety of concave and plane reflection gratings. Used in first or higher orders.	600–3600/mm	2° to 22°	Al protected by MgF_2 recommended for gratings in this region.
Ultraviolet 2000 Å to 4000 Å Visible 4000 Å to 7000 Å Near infrared 0.7 to 2.0 microns	Over 500 sizes of plane or concave reflection gratings used in first or higher orders. Also a selection of plane transmission gratings.	300–2160/mm	3° to 64°	Plane for photoelectric spectrometers; concave for photographic recording of spectra.
Infrared 1.0 to 7.5 microns	Complete range of sizes and types of plane gratings. First order.	20–600/mm	5° to 30°	Plane gratings with aluminum coating.
Far infrared 40 to 1000 microns	Plane gratings ruled to order in aluminum metal. Any desired size up to 330 × 360 mm.	0.8 to 12/mm	15° to 31°	SiO coated . . . used with plane filter gratings.

*Courtesy Bausch and Lomb, Inc., 1968.

REPLICAS VS ORIGINALS

Most gratings sold for laboratory use are replicas, rather than originals. Many replicas can be made from a single original, and they can be equally good. A sub-master of expoxy resin is made from the original grating, and a thin replica is made from this submaster. It is then mounted on a glass or fused silica blank just as the original, and a reflecting coating of aluminum deposited on it. To all appearances the replica and original are the same, and none of the tests described can distinguish one from the other if the replica is properly made.

REPRESENTATIVE LIST OF RESEARCH GRATINGS

One grating manufacturer, Bausch and Lomb, has prepared a chart showing the wide range of research grade diffraction gratings obtainable, together with comments. This is reproduced in Table 2.

3

Aberrations

Only a general description of aberrations is given, because each mounting has its own characteristic ones.

Astigmatism is present when a point on the slit is imaged as a line at the receptor, rather than as a corresponding point. The aberration may be troublesome because it results in a loss of intensity when a photographic plate is used, unless a long length of entrance slit is illuminated. Also, it makes impossible the use of slit masks to delineate spectra, and requires the use of occulters or deckers at the plate. The resolving power may also be degraded.

Coma is present when a point on the slit is imaged as a comet-like shape at the receptor. If the comet tail is directed along the spectral line not much loss of resolution results.

Curvature of the field is present when the focal curve is not a straight line. This aberration, so troublesome in ordinary image forming optics is not often a serious problem in grating spectrographs. The focal curve can be experimentally determined, and the plateholder made to conform to this curve. In some mountings the shape of the focal curve changes as the angle of the grating is changed, and then of course the plateholder must incorporate some means of changing its curvature. Most mountings are designed to have a fixed focal curve for the entire range of grating angles which are used.

Distortion is always present, and results in curved images of the slit. In many spectrographs it is too small to be noticeable, except when accurate wavelength measurements are being made.

Grating Mountings

CONCAVE

GENERAL DISCUSSION

A concave grating is a reflection grating whose grooves are ruled on a concave spherical mirror, with equal spacings along a chord of the surface. The purpose of ruling a grating in such a fashion is to utilize the focal properties of the mirror while at the same time utilizing the diffracting properties of the grooves. H. A. Rowland discovered that such a grating properly mounted can simultaneously disperse light and focus it into a spectrum. The Rowland circle has a diameter equal to the radius of curvature of the grating blank, and is constructed with a tangent coincident with a tangent to the grating surface. The grating grooves are placed perpendicularly to the plane of the circle. Light passing through a slit located anywhere on the circle and striking the grating is diffracted and brought to a sharp focus as spectral lines at points on the circle corresponding to the wavelengths present in the source. The positions of the images are given by the usual equation, $m\lambda = a(\sin\theta + \sin\phi)$. A concave grating can also be illuminated in collimated light; the focal curve is then no longer a circle.

Research gratings are usually of 1, 1.5, 2, 3, 6.4, or 10.7 meter radii of curvature, and are popularly called "one-meter concave gratings," and so forth. Ruled widths are available up to 20 cm, although the quality of the larger gratings is often not as good as the smaller ones, since the longer a time an engine rules, the more difficult it is to maintain the same degree of accuracy in ruling. Random errors are more likely to occur, and al-

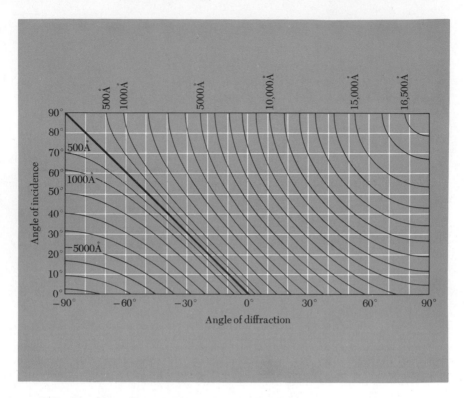

FIG. 12. Nomogram showing the relation between wavelength and angles of incidence and diffraction for a concave diffraction grating having 1200 grooves/mm. [After H. G. Beutler, *J. Opt. Soc. Am.* **35,** 311 (1945).]

though the resolution is greater, it is a smaller percentage of the theoretical value.   Grating spacings range from 1/3600 to 1/300 mm.

The spectral ranges for various angles of incidence and diffraction are shown in Fig. 12.   The figure can be appropriately scaled for gratings of spacings other than the one illustrated.

PASCHEN–RUNGE

The Paschen–Runge mounting of a concave grating is an installation making use of almost the entire Rowland circle, as shown in Fig. 13.   The angle of incidence ϕ is always positive; the angle of diffraction θ is positive when it is on the same side of the normal as the angle of incidence, and negative when it is on the opposite side.   It is set up in a light-tight room, with slit, grating, and plate holder mounted on concrete piers in order to minimize vibrations.   A representative grating has a radius of 6.4 m, is ruled with 1200 grooves/mm, blazed at 7500 Å in the first order, and has 15 cm of ruled width.   Let us choose this grating as an example with the angle

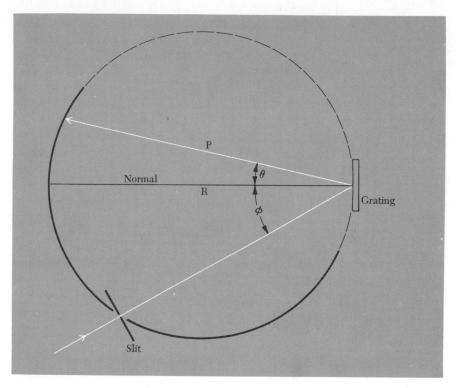

FIG. 13. Schematic diagram of Paschen–Runge mounting. (As shown here, θ is negative.)

of incidence 40 degrees. The wavelength region in a practical mounting extends from 2000 Å at $-24°$ to 12,000 Å at 53° in the first order. It is practical to use the first, second, and third orders; ruling errors usually preclude the use of higher ones. The ranges of angles of incidence and diffraction ordinarily used are shown in Fig. 14.

The theoretical resolving power $R = mN$, or 150 mm $\times$ 1200 gr/mm $= 180,000$ in the first order. In practice the resolving power is usually only about one-half the theoretical value, owing to ruling errors and aberrations; a round number of 100,000 is typical.

The plate factor is given by $\dfrac{d\lambda}{dl} = \dfrac{\lambda \cos \theta}{P(\sin \theta + \sin \phi)} = \dfrac{a \cos \theta}{mP}$. However, it must be remembered that the Rowland circle (and hence the receptor) is not perpendicular to the path of light from the grating, nor is the projection distance P a constant. With the appropriate substitutions, $\dfrac{d\lambda}{dl} = \dfrac{a \cos \theta}{mR}$. In the example chosen, the ratio a/mR has a value of 1.3 Å/mm in the first order, 0.66 Å/mm in the second order, and 0.42 Å/mm in the third order.

The free spectral range given approximately by $F_\lambda = \lambda/m$ is large, since the orders normally used are the first, second, and third. It is more useful and accurate to think of what wavelengths appear at the same angle, rather than what wavelength range can be observed without overlapping. For example, 2500 Å in the third order, 3750 Å in the second order, and 7500 Å in the first order appear at the same angle of diffraction. The two undesired orders can be eliminated by the use of filters or a predisperser, as described in Chapter 5.

The aperture ratio of a concave grating spectrograph is small when compared to most optical instruments. The important dimensions are the projected width of the grating and the distance from it to the receptor. In our example, with a 15 cm grating and a radius of curvature of 6.4 m, the ratio is

$$R \cos \theta / W \cos \theta = R/W = 43.$$

The region of the blaze is somewhat larger than the approximate theoretical value $\Delta\lambda = \pm\lambda/2m$, owing to both imperfections and change in

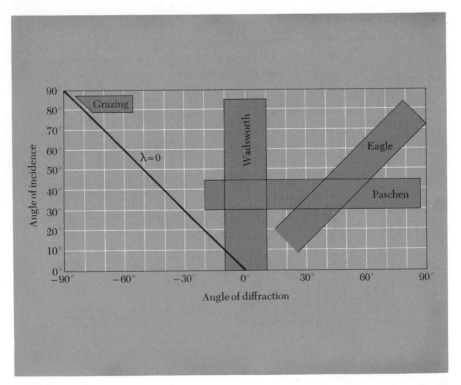

FIG. 14. Angular range covered by various grating mountings. [After H. G. Beutler, *J. Opt. Soc. Am.* **35**, 311 (1945).]

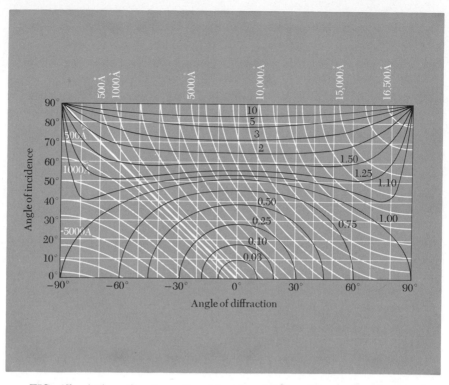

FIG. 15. Astigmatism in units of length of the grating grooves for a point source at the slit. [After H. G. Beutler, *J. Opt. Soc. Am.* **35**, 311 (1945).]

groove shape across the face of the grating. The ruling diamond is held in a fixed orientation throughout the ruling process, but the slope of the blank surface changes continuously from one end of the grating to the other. Therefore, the grooves are not identical in shape, and not all are equally effective in diffracting light into a given order. The maximum blaze intensity is reduced, and the light is spread into more orders than would be expected for identically shaped grooves. For example, using just the theoretical value, light is diffracted with appreciable intensity from 3750 Å to 11,250 Å. In practice, spectra can be photographed from 2200 Å to 12,000 Å.

In this instrument, the focussing properties are provided for by the spherical mirror although the direction of light is controlled by the diffraction. Consequently, the horizontal and vertical foci are at different distances from the grating, and the amount of astigmatism can be extreme. The length of the image of a point on the slit is shown in Fig. 15. If full intensity is desired, a length of slit approximately equal to the length of the astigmatic image must be illuminated. Further, the slit images are slightly curved, although the curvature is often not noticable.

EAGLE

The Eagle mounting has nearly equal angles of incidence and diffraction, as shown in Fig. 14.   A schematic drawing is shown in Fig. 16.   Provision is made for rotating the grating in order to change spectral regions. When the grating is rotated, the plate holder is also rotated and the distance between slit and grating is changed, in order to keep the slit and plate holder on the Rowland circle.   The entire instrument can be mounted in a relatively small box, as compared to the Paschen mounting.   Of course, only a small spectral range can be covered at a single setting.   In all other respects the Eagle mounting is similar to the Paschen.

GRAZING INCIDENCE

The grazing incidence mounting is especially useful in the vacuum ultraviolet region of the spectrum.   The angles of incidence and diffraction are shown in Fig. 14, and a schematic drawing is shown in Fig. 17.   The reflectivity of metals decreases sharply in the short wavelength region of the spectrum, but when very small angles of incidence are used, the reflectance increases somewhat.   A LiF overcoating will increase the reflectance

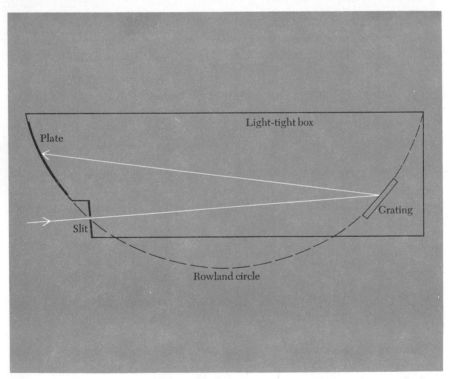

FIG. 16. Schematic drawing of an Eagle mounting.

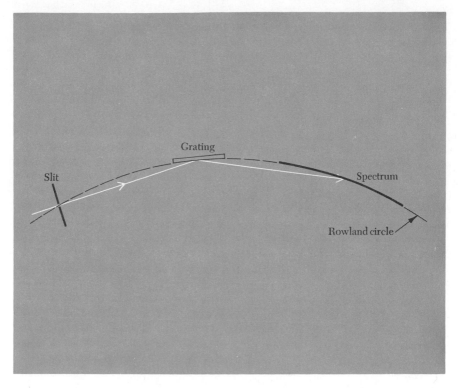

FIG. 17.   Schematic drawing of a grazing incidence mounting.

substantially.   At these small angles, the spectrograph is very small as compared to the Paschen mounting, and hence is well adapted to placement in a vacuum chamber.

Some of the drawbacks are evident; astigmatism is extreme, focussing is difficult because the focal distance from the grating to the circle is a rapidly varying function of angle, the circumference of the circle is inclined at a steep angle to the incident and diffracted light paths, and the plate factor varies rapidly with wavelength.   There is another more serious restriction on the use of a grazing incidence mounting.   The inherent aberrations impose a limit on the maximum size of grating that can be used without a loss of resolution.   The details of the calculation provide little physical insight into the reason for the limitation, and hence will not be reproduced here.   For simplicity, the grating is considered to be circular in aperture, with radius ρ_0; R, a and m have the usual significations of radius of curvature, groove spacing, and order.   The optimum aperture of grating as a function of angles of incidence and diffraction is shown in Fig. 18. As an example, let us choose a grating with 1200 grooves/mm, 3 m radius of curvature, used at an angle of incidence of 85°, in the wavelength range from 100 to 1000 Å. For 100 Å, the angle of diffraction is $-80°$, and the plate factor 0.48 Å/mm. For 1000 Å, the corresponding figures are $-61°$ and 1.35 Å/mm. The optimum size of the aperture of the grating has a radius of 1.2 cm at 100 Å, and 2.5 cm at 1000 Å. A grating larger than this size may still be useful, but the resolution will not be proportionably greater. The references give further details. This size limitation is not so strict for other mountings, and available gratings are not yet large enough to be the optimum size.

WADSWORTH

In the Wadsworth mounting, collimated light is incident upon the grating, as shown in Fig. 19. The spectra are observed at small angles centered on the normal to the grating ($\theta = 0°$). The astigmatism is eliminated in this mounting, and hence slit masks can be used to delineate spectra. Also, the mounting can be used in conjunction with externally mounted dispersing optical elements such as the Fabry–Perot interferometer. The many subtleties of the mounting are adequately described in Beutler's

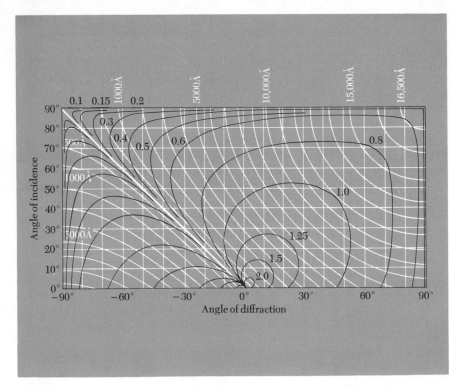

FIG. 18. Optimum radius of a grating in units of $\sqrt[4]{\dfrac{2R^3a}{m}}$. [After **H. G.** Beutler, *J. Opt. Soc. Am.* **35**, 311 (1945).]

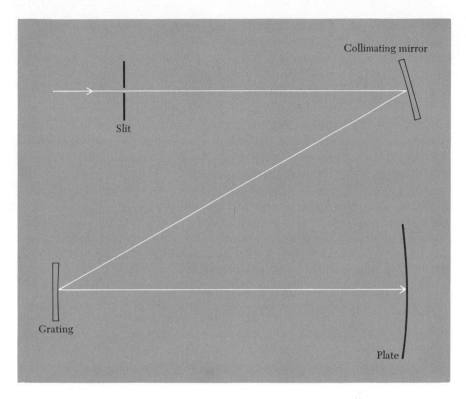

FIG. 19. Schematic drawing of a Wadsworth mounting.

article. Again, for comparison, let us consider the same grating chosen earlier in the Paschen mounting, a grating with 1200 gr/mm, a radius of 6.4 m, and 15 cm of ruled width. If it is used at an angle of incidence of 40°, the spectrum from 4000 to 6800 Å appears in a range ±10° from the normal, in the first order. The focal curve is no longer the Rowland circle, but is given by

$$r' = R \cos^2 \theta / (\cos \theta + \cos \varphi).$$

In our example, at the normal to the grating ($\theta = 0°$, 5400 Å) the focal distance is 3.62 m. The theoretical resolving power is the same as in the Paschen mounting, 180,000 in the first order. The plate factor is larger (smaller dispersion) by the ratio R/P, or by a factor of 1.7. The free spectral range remains the same. The aperture ratio is increased by the same factor, and the speed by the square, or a factor of three.

To cover a different region of the spectrum higher orders can be used, or the grating can be rotated, the shape of the plateholder changed, and the projection distance changed. These adjustments make the use of the Wadsworth mounting difficult in many circumstances.

SEYA–NAMIOKA

The Seya–Namioka spectrograph is a compact monochromator for the vacuum ultraviolet, in which the spectrum is scanned by rotating the grating about a vertical axis through the grating center. No simplified description of the optical properties of this mounting is possible. Detailed calculations show that aberrations are minimized for an angle of about 70 degrees between lines connecting the center of the grating with the entrance and exit slits. An equivalent statement is that the sum of the angle of incidence and the negative of the angle of diffraction is constant, and equal to 70 degrees. The slits and grating are not placed on the Rowland circle.

This mounting is compact and stable, since only one motion—rotation of the grating—is needed to scan. Also, the direction of the diffracted light is constant for all wavelengths. It has the disadvantages of large astigmatism (never less than 0.66 the length of the grooves) and multiple or asymmetric lines which are troublesome except for small grating widths. Hence, high resolving power is not obtainable.

PLANE GRATING MOUNTINGS

Plane gratings are simply diffraction gratings ruled on plane surfaces, as described earlier. It is somewhat easier to rule nearly perfect plane as compared to concave gratings, and nearly all fundamental research in ruling has been conducted with them. When used in spectrographs, they have the disadvantage that image forming optics separate from the grating must be used, with a consequent introduction of more complexity to the instrument. They are more accurately ruled, however, and the practical resolving power of a plane grating instrument may reach the theoretical limit. Also, because the projection distance is not a function of wavelength or the angle of incidence of light on the grating, the resulting instrument is a more stable and accurate one, particularly for wavelength measurements. Further, the spectra are stigmatic, or nearly so. Plane gratings are available in sizes ranging up to 30.5 cm in ruled width, with groove spacings from 1/25 mm to 1/1200 mm. Larger gratings do exist, but they are montages—several smaller gratings mounted together—and do not show greater resolution than a single, smaller grating. An increase of speed can be gained through use of a montage.

CZERNY–TURNER

There are many types of mountings, but a representative one is the Czerny–Turner mounting, in which there are a collimating mirror and a camera mirror, as shown in Fig. 20. It should be evident that there are many variations of the mounting; the mirrors need not have the same focal length, the angles need not be equal, the grating can be placed at other distances

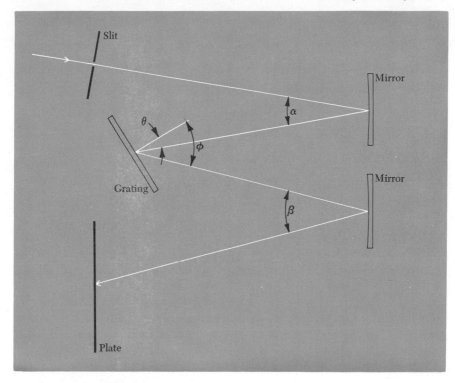

FIG. 20. Schematic drawing of a Czerny–Turner mounting.

from the mirrors, both mirrors can be combined in one, the slit, grating, and receptor can be arranged vertically instead of horizontally, etc. Each has some particular advantage, but the Czerny–Turner is illustrative of the general characteristics of plane grating spectrographs. A very common focal length is 3.4 m for both mirrors. Some of the larger research installations reach 15 meter focal lengths. For an example, consider a mounting as shown in Fig. 20, with a grating having 15 cm of ruled width, a reciprocal spacing of 300 grooves/mm, blazed at 64° (blazed at 6 microns in the first order), with both mirrors of 3.4 m focal length.

The approximate equation for the resolving power is

$$R = \frac{2W \sin \theta}{\lambda} = 2.7 \times 10^9/\lambda$$

with the wavelength in Å. At 5000 Å in the 12th order the resolving power is 540,000. Correspondingly, the resolving limit is approximately $1/(2W \sin \theta) = 0.037$ cm^{-1} at all wavelengths.

The approximate equation for the plate factor (for equal angles of incidence and diffraction) is

$$\frac{d\lambda}{dl} = \frac{\lambda}{2P \tan \theta} = 7.35 \times 10^{-5}\lambda \text{\AA/mm}$$

near the blaze angle, with the wavelength in Å. At 5000 Å in the 12th order,

$$\frac{d\lambda}{dl} = 0.37\text{\AA/mm},$$

or equivalently

$$\frac{d\sigma}{dl} = 1.5 \text{ cm}^{-1}/\text{mm}.$$

The free spectral range is given approximately by $\pm\lambda/2m$ or $\pm\sigma/2m$. At 5000 Å, the values are 210 Å and 85 cm^{-1}.

The aperture ratio is $P/(W \cos \theta) = 45$. This figure is about the same as for the equivalent concave grating, yet in practice the plane grating instrument is much faster. For one thing, all grooves have the same shape (except for changes resulting from diamond wear or chipping during ruling), so the diffracted wavefront is more uniform in amplitude than for a concave grating. For another the groove spacing is usually much larger, and hence the blaze efficiency is greater, as mentioned earlier. A third reason for increased speed is the lack of astigmatism. An increase in speed by a factor of 10 over the equivalent concave grating is not unusual.

The grating must be used at or near the blaze angle in order to make best use of its groove shape. A maximum range of about 10 degrees is an acceptable variation in angle for a grating ruled with 300 gr/mm, although there is always some light in the spectra at smaller angles. At larger angles, the intensity drops very rapidly. When determining what angles of incidence and diffraction to use, it is helpful to think in terms of $m\lambda$ rather than λ alone. In the example chosen, the grating is useful at angles from 56 to 67 degrees, or over the range 54,700 to 61,100 order-angstroms. A wavelength of 3000 Å can be observed in the 19th and 20th orders, 5000 Å in the 11th and 12th orders, and 7000 Å in the 8th order. Observation of longer wavelengths may present a problem; 9000 Å in the 6th order falls at a smaller angle than the limit set, while in the 7th order it falls at a larger angle. The only procedure is to use the smaller angle, and accept a diminished intensity because the observation angle is far from the blaze. A short calculation will show that a larger groove spacing would permit the observation of longer wavelengths within the useful range of blaze angles.

The Czerny–Turner mounting has several optical aberrations which can change the appearance of the spectrum and may degrade the resolving power. The images of the slit are curved, sometimes severely. No special difficulty is encountered as a result of this curvature, but care must be used when making wavelength measurements. Spherical aberration

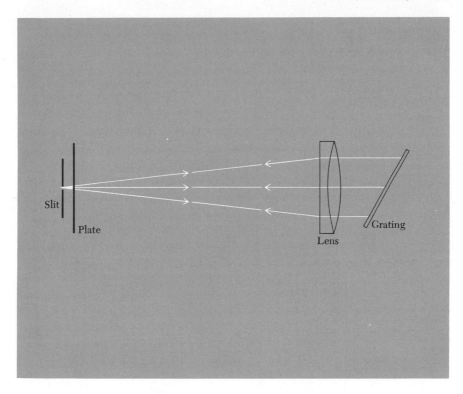

FIG. 21. Schematic drawing of a Littrow mounted plane grating.

is present in all instruments which use large aperture spherical mirrors. In the example chosen, the aperture is small enough that the aberration is not a problem if the angles α and β are kept as small as physically possible. Coma is also present, although the tail is directed along the spectrum line and the line is only slightly widened. To compensate for coma, the relative focal lengths of the collimating and camera mirrors can be adjusted appropriately, or the grating can be placed asymmetrically between the mirrors. Both methods are utilized in commercial instruments.

LITTROW

The most compact of all plane grating instruments is the Littrow mounting, shown in Fig. 21. Distinguishing features are the use of a single focussing element, and equal angles of incidence and diffraction. A number of variations in location of slit and receptor are possible. The mounting is seldom used except for a specific restricted region of the spectrum, since it is difficult to obtain large well-corrected lenses. On the other hand the mounting is well-adapted to scanning the spectrum by rotation of the grating. The spectroscopic properties are similar to those of the Czerny–Turner mounting discussed earlier.

MULTIPLE PASS

Attempts have been made to improve the resolution of grating instruments by using multiple diffraction from a single grating.   In these instruments, light is reflected back to the grating one or more times after the first diffraction.   The resolution of the instrument is thereby increased, as is the dispersion.   However, the speed of the instrument is smaller because of the extra reflection from the grating (as well as the mirrors), and numerous secondary images are formed, which can be quite troublesome.   It is usually better to utilize a larger single pass instrument, than to work with a multiple pass one, although some very successful multiple pass instruments have been constructed and are in use.

ECHELLE

The development of plane gratings and their general usage in high resolution spectrographs began in the late 1940's, and was started primarily by the introduction of the echelle spectrograph by George R. Harrison.   Prior to this time, it was common to use concave diffraction gratings in the first, second, and third orders to give resolutions up to perhaps 200,000.   For

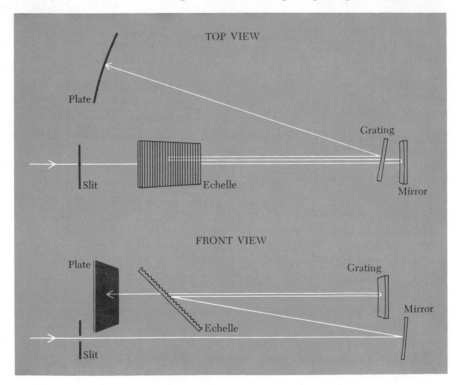

FIG. 22. Schematic drawing of an Echelle spectrograph.

$\lambda \longrightarrow$

Hg 5461Å

Order

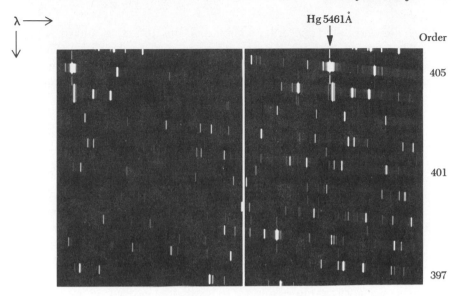

405

401

397

FIG. 23. Echellegram of the thorium spectrum.

higher resolution, it was necessary to go to interferometers or echelons, operating in orders of interference of about 40,000.   Harrison's idea was to make use of some of the known but little used properties of plane gratings: use at large angles of interference and diffraction, to take advantage of consequent high resolution and dispersion; use of a coarse spacing intermediate between then current spacings of gratings (1/600 to 1/1200 mm) and interferometers (1 cm), resulting in an intermediate value of the free spectral range.   All wavelengths appear at about the same angle of diffraction, and a well-shaped groove with a good blaze at all wavelengths is possible.   Harrison proposed crossing this echelle grating with an instrument of moderate dispersion, and displaying a two-dimensional spectrum much as in an interferometer-spectrograph combination, but now encompassing the entire spectrum rather than a small portion. A schematic of a successful instrument is shown in Fig. 22, and might be called an internal Wadsworth mounting of an echelle grating.   Light from a horizontal slit is collimated by the first mirror, and sent to the echelle, which is placed nearly over the slit, with its grooves horizontal.   Light diffracted from the echelle is sent to a concave diffraction grating placed above the collimating mirror, with its grooves vertical.   The light is then diffracted to the photographic plate holder placed to one side of the echelle.   Since the dispersion of the grating is perpendicular to that of the echelle, a two-dimensional spectrum is produced.

The choice of parameters is wide, but certain relations must be preserved if the spectrum is to be useful.   The primary dispersing element is the echelle, and its characteristics determine the resolution and dispersion of the instrument (the echelle grooves are parallel to the slit).   The free spectral range of the echelle (called an echelle cycle) must be small enough to fit into the smaller dimension of the photographic plate.   The concave grating must have enough dispersion to separate adjacent echelle cycles for some minimal slit length, and its free spectral range must fit into the larger dimension of the photographic plate.   In one particular instrument, the echelle has 12 cm of ruled width with 8 grooves/mm blazed at 63°, and the concave grating a radius of 6.4 m with a reciprocal spacing of 600 grooves/mm.

The spectrum appears from 3500 Å to 7000 Å in the first order of the grating, but in orders 636 to 318 of the echelle.   Wavelengths shorter than 3500 Å appear in the second order of the grating.   It is possible to obtain the spectrum from the short wavelength limit of transmission through air to 7000 Å on a single 5 × 50 cm photographic plate.   The shorter wavelengths belonging to echelle cycles in the second order of the concave grating are easily separable from those in the first order because they are one-half as long, and appear with one-half the spacing.   Unless the spectrum is very dense, there are not enough coincidences in spectrum lines to require separation of wavelengths by a filter.   An echelle spectrogram is shown in Fig. 23.   In spite of the compact nature of the spectrum display, this instrument has a resolution greater than 200,000 at 5000 Å, and a plate factor of 0.24 Å/mm at 3500 Å and 0.47 Å/mm at 7000 Å. In actual use, it is superior in speed and resolving power to a 15 cm, 10.7 m radius of curvature concave grating.   It has further advantages of being a more stable instrument, and more compact.

There are several unique problems in the adjustment and usage of the echelle spectrograph, the most prominent of which is the psychological difficulty of reducing data presented in a two-dimensional pattern of spectral lines.   At first glance an echellegram of a complex spectrum appears hopelessly jumbled, and impossible to measure.   With a little patience and experience, this problem disappears.   Further, the slit length must be made fairly short, perhaps two millimeters, resulting in an apparent accentuation of any optical aberrations such as astigmatism and coma. Actual data reduction is not measurably affected by short spectrum lines. However, to help overcome these problems, echelles have been designed with smaller spacings then the one just discussed, having larger free spectral ranges.   The cross dispersion has also been increased to allow longer slit lengths.   Most of these instruments require more than one setting of the grating (cross dispersion) to cover the entire spectrum.   The high resolution and dispersion features of the echelle are preserved.

An echelle spectrograph incorporates at least one additional reflection over the usual spectrograph, and consequently suffers an additional loss of light.   This loss becomes greater as wavelengths become shorter.   Another significant disadvantage of an echelle is the necessity of maintaining good focus over a three-dimensional region.   For example, the Wadsworth mounting previously described is not completely free from aberrations, particularly astigmatism.   Also, coma becomes a problem.   Consequently, the sharp focus is degraded toward the corners of the plate.   As a general rule, however, the increased resolution and dispersion are enough to allow some aberrations to be present, without degrading the performance down to the level of ordinary mountings giving much lower resolving power and dispersion.

Miscellaneous Topics

SLIT ILLUMINATION AND WIDTH

The illumination of the slit and the slit width determine how much light enters the spectrograph and produces a useful diffraction pattern. The solid angle Ω of light accepted by the spectrograph is determined by the size of the dispersing element and the focal length of the collimating lens or mirror. This solid angle must be geometrically filled with light from the source, at the very least, in order to utilize the instrument properly. If an image of the source is focussed on the slit, the solid angle of light from the condensing lens Ω_c must be at least as large as the solid angle of the spectrograph, as shown in Fig. 24. No loss of light occurs if Ω_c is larger than Ω, providing that the image of the source is at least as large as the slit. (To see this, note that the brightness of a source and its images are always the same, regardless of the kind of image forming optics, and that the useful radiation entering the spectrograph is the product of the slit area and the solid angle, a constant.)

If an extended source is placed near the slit, the solid angle subtended by it at the slit should again be not less than the solid angle of the spectrograph. (Often a lens which images the source on the spectrograph collimating lens is placed at the slit.) It is not a trivial matter to insure that the spectrograph solid angle is filled with light. An easy way to tell whether this condition is fulfilled is to set up a source with a bright spectrum line as in Fig. 24. Put your eye in place of the receptor, and look at the face of the grating. If the entire face is illuminated, the spectrograph is filled

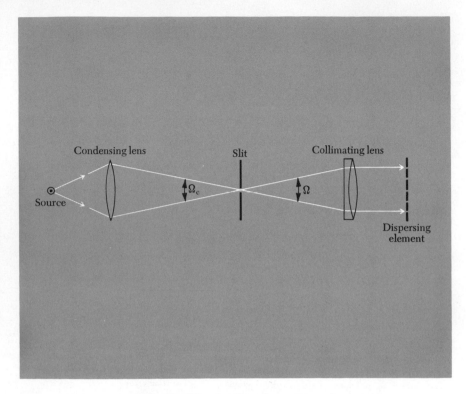

FIG. 24. Illumination of spectrograph.

with light as it should be. The two cases just discussed are called *partially coherent* illumination of the spectrograph. *Incoherent* illumination occurs when an extended source is placed right at the slit—a condition seldom attainable in practice. *Coherent* illumination occurs when a small source is placed a large distance away from the slit, with no condensing lenses in between.

The slit width to be used depends on the method of illumination, the fraction of theoretical resolution desired, and the throughput required. It will be assumed that the illumination is partially coherent and that the light from the source fills or overfills the spectrograph solid angle. Then the resolving power, intensity at the peak of a spectral line, and the slit width are interrelated as shown in Table 3. The unit of slit width is taken as the wavelength of light times the aperture ratio, $\lambda f/D$. For example, consider an $f/25$ spectrograph. At 5000 Å, a slit width of 12.5 microns will allow a resolution 90% of theoretical with a relative intensity of 0.61; a width of 25 microns 52% with a relative intensity of 0.87. The optimal slit width, defined as that width which produces a maximum of the product of resolution and intensity, is approximately 1.25 $\lambda f/D$, or 15 microns in this case, for which the resolution is 0.79 of theoretical and the relative intensity 0.75.   The values of slit widths and relative intensities given in Table 3 are calculated for monochromatic light incident on the spectrograph.   Such a circumstance never occurs in practice, since radiation in any given spectrum has some minimum wavelength spread about each spectral line; this spread determines the minimum wavelength separation that can be resolved irrespective of the spectrograph.   However, if the spectrograph instrumental width is much larger than the intrinsic spectral line width (the *unusual* case), the figures given accurately reflect practical limits.   When the instrumental width and spectral widths are about the same, the tabular values must be modified.   Rather than make this modification mathematically, the tabular values are taken as a first approximation, and suitable changes experimentally determined.   A useful method is to observe a strong spectral line while varying the slit width. Relative intensities are impossible to evaluate in this manner, but changes in resolving power are strikingly easy to observe.

TABLE 3
Spectrograph Slit Width, Intensity, and Resolving Power

Slit Width w $(\lambda f/D)$	Relative Intensity I/I_0	Fraction of Theoretical Resolving Power R/R_0
0.0	0.00	1.00
0.5	0.22	0.98
1.0	0.61	0.90
1.25	0.75	0.79
1.5	0.82	0.71
2.0	0.87	0.52
3.0	0.89	0.34
4.0	0.93	0.26
6.0	0.95	0.17
8.0	0.96	0.13

It should be evident that the figures also depend on having a perfect diffraction grating.   Some well-ruled plane gratings reach this standard and show full theoretical resolving power.   Most concave gratings fall far below this standard, and slit widths given in the table can often be doubled without seriously degrading the actual resolving power.

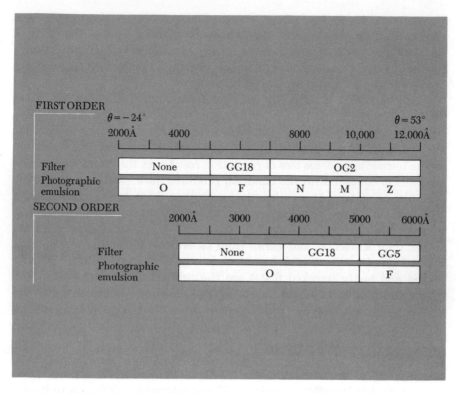

FIG. 25. Overlapping orders in a concave grating spectrograph.

$$F_\sigma \simeq \frac{1}{2a \sin \theta} \quad \text{or} \quad F_\lambda = \frac{\lambda^2}{2a \sin \theta}$$

as before. Let us use the former example of a 300 groove/mm plane grating in a Czerny–Turner mounting at 64 degrees. The free spectral range is 1667 cm^{-1} or 1.67×10^{-5} λ^2 Å. A tunable band pass filter is needed. If only a limited range of the spectrum is to be observed, a few filters will suffice. Usually this is not the case, and a pre-disperser is required. A pre-disperser is merely an instrument of low dispersion which forms a spectrum on the slit of the higher dispersion instrument. It can take many forms, but the two most useful ones are band pass interference filters combined with glass filters, and small low-dispersion spectrographs between the source and slit.

One example is a small prism spectrograph set up as shown in Fig. 26. The purity of the spectrum formed at the slit of the spectrograph is determined by the width of the pre-slit. The usual procedure is to set the pre-slit just wide enough to encompass one free spectral range at a single

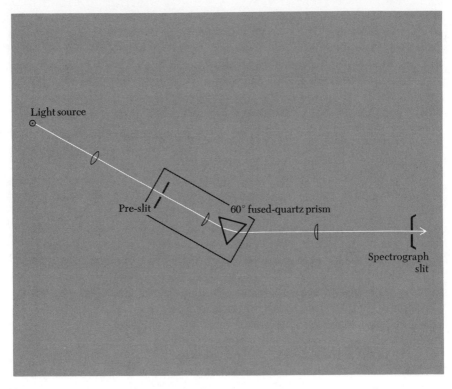

FIG. 26. Prism predisperser.

point in the spectrum, centered at the desired wavelength. Further practical suggestions are given in the references.

WAVELENGTH MEASUREMENT

Accurate wavelength measurements are essential to many spectroscopic problems. The most accurate ones are made with interferometry, but good wavelengths, accurate throughout the spectrum to ± 0.015 cm^{-1} or slightly better, can be made with reasonable care on a 3 m Czerny–Turner mounting. It is essential to have a grating which produces symmetrical spectrum lines, and a standard comparison spectrum whose wavelengths are accurately known. Most work today uses thorium lines as standard wavelengths. Really good secondary and tertiary standards have not yet been universally accepted, but good measurements can be made using thorium. The best procedure is to take two spectra overlapping to a small extent—one of the standard, and one of the unknown. Comparator measurements should be made running down the region where the standard and the unknown overlap. Some blending will occur, and some unknown lines cannot be measured accurately in this fashion. If the comparator is equipped with a good cross slide, however, the plate can be moved out of the region of overlap, and a measurement taken. This measurement cannot be considered as accurate as the others, owing to the presence of aberrations such as distortion. Every spectrograph has some aberrations, and introduces errors by this procedure. From the known wavelengths of the thorium standards, a calibration curve can be calculated (by computer, or by hand if necessary), and the unknown wavelengths interpolated. Each experimenter will work up his own procedure; the purpose here is to point out that the only accurate way is to have the standard spectrum overlap the unknown.

RELATIVE INTENSITY MEASUREMENT

It is often useful to have an indication of relative intensities of spectral lines. For this purpose, the best method is to scan the lines with a linear response detector, such as a photomultiplier. For spectra recorded photographically, only a rough calibration can be made of intensity vs density, but even so it may be good enough for most purposes. In this case, a calibrated step weakener or neutral density filters can be placed at the slit of a stigmatic spectrograph, or immediately in front of the plate in an astigmatic one. Step weakeners are made of metal evaporated onto a thin quartz plate in strips of varying densities. They are calibrated with a linear response detector at the wavelengths for which they are to be used.

If it is inconvenient to expose the spectrum through the step weakener, a white light source with the appropriate wavelength filter can be used for plate calibration. Here, however, care must be taken to keep the exposure time approximately the same as that for the spectrum, in order to avoid the effects of reciprocity failure. Other methods are amply covered in references 2 and 3 of Chapter 1.

CONCAVE GRATING SPECTROGRAPH ADJUSTMENT

The initial setting-up and adjustment of a concave grating requires considerable time and effort if the best possible performance is to be achieved. The slit holder should be adjustable in height, and have provisions for leveling. It should be mounted on a slide for motion along a line joining the slit and grating. The grating holder should also be adjustable in height, and allow rotation of the grating about three mutually perpendicular axes. In a Paschen mounting, which is the only one for which adjustments are discussed, the track should be adjustable in height as well as radius of curvature. Because the mounting is astigmatic, occulters or deckers are needed directly in front of the track, in order to photograph spectrum lines of a limited length.

To start with, set the slit, grating, and track in approximately the correct positions. Next, set them at the same height, using a transit for this purpose. Then point the slit base directly at the grating, using the following procedure. Set up a mercury arc in front of the slit, but some distance away from it. Adjust its position so that light passing through the slit falls on the center of the grating. Then use a lens to focus an image of the arc on the slit; light should now cover the entire face of the grating. At the proper position on the track, observe the mercury green line in the first order, with a magnifier. Now move the slit assembly toward and away from the grating while observing the green line. Adjust the direction of the slit base until the image of the line does not move sideways as the slit is moved. Of course the line will not remain in focus, but the center of the image should stay in the same place on the track. After this adjustment, the base can be fixed in place.

The slit should next be made vertical. Set up a light source as before, focussed on the slit. Behind the slit (on the grating side) place a lens so that it forms a greatly enlarged image of the slit on a white card placed just in front of the grating. Place a plumb bob in front of the card, and

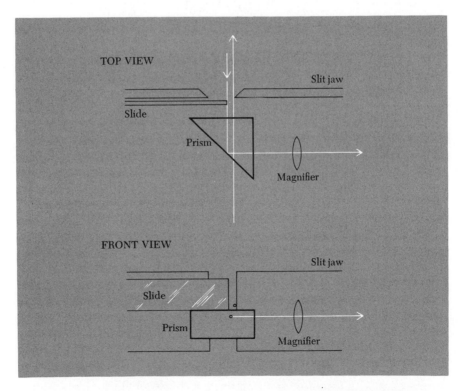

FIG. 27. Optical arrangement for setting grating angle and slit position.

adjust its position until its shadow bisects the image of the slit.   (The tip of the bob can be immersed in a beaker of water, to damp out oscillations.) Any deviation of the slit from vertical can easily be seen and should be corrected.

The next steps are to set the grating at the proper angle, and to put the slit on the Rowland circle.   The angle of incidence is determined by the groove spacing, blaze, and desired wavelength region of operation.   For example, suppose that the grating has 1200 gr/mm and is blazed at 7500 Å in the first order, and that the spectrum up to 11,000 Å is to be observed (11,000 Å at the slit, and also 5500 Å in the second order).   The angle of incidence should be approximately 41 degrees.   Set up a source which has a spectrum line at or near 5500 Å (neon, for example), and arrange a prism and eyepiece so that the wavelength diffracted directly back through the slit can be observed, as shown in Fig. 27.   Next rotate the grating about a vertical axis until the desired wavelength appears in the field of view, and adjust the slit position on its base until both the spectrum line and the slit jaws appear in sharp focus.   The slit is now set on the Rowland circle and can be locked in place.

The grating grooves should now be adjusted to the vertical position. Make a further preliminary setting of the track position by eye, using prominent lines in the mercury spectrum.   Then illuminate the slit and grating with white light, so that a continuous spectrum is formed all around the track.   Rotate the grating about one horizontal axis until the spectrum is at the center of the track at the diffraction angle of zero degrees (normal to the grating).   Rotate the grating about the other horizontal axis until the spectrum is at the center of the track at angles of diffraction as far from the normal as possible.   Several readjustments of these rotations are usually required.   At this point it is a good idea to recheck to see if the slit is still on the Rowland circle.

The setting of the track curvature and distance is the next adjustment to be made.   Take three slant-focus plates at widely spaced sections of the spectrum, using a thorium lamp as a source.   (A small cassette can be made to clamp to the tracks.   It should hold a section of plate about 10 cm long inclined at an angle of 45° to the vertical direction.)   Set the track to the position of best focus at these three places.   To determine the position of the rest of the circle, set up a long lever arm pivoted at the center of the Rowland circle determined by the three reference points just found.   Swing the arm around, and adjust the track to the circle as accurately as possible. A dial indicator fastened to the end of the arm is a useful measuring instrument.   The tolerance should be within ±0.5 mm, and preferably smaller.

The next step is to take focus plates at regular intervals, to see if any slight readjustments are necessary.   If a choice must be made between sharp lines with satellites or fuzzy lines with no satellites, it is better to choose the sharp lines and mask the grating to eliminate the satellites, if at all possible.

An effective way to determine the necessary masking is to observe the mercury green line hyperfine structure while another person systematically covers portions of the grating.   It should be noted that a person cannot stand near the grating while masking it.   Body heat causes air currents which seriously distort the image.

Now illuminate a very short section of the slit and take spectrograms of the entire length of the astigmatic images.   Mask the ends of the grooves of the grating as much as is necessary to put the spectrum lines in equally sharp focus from top to bottom.   If the grating is masked asymmetrically in either direction, the center of the grating should be reset, and the above procedures repeated.   As a final test, it is worthwhile to take spectrograms of thorium over the entire wavelength region to be used.   These can subsequently be used for maps to identify unknown wavelengths.

PLANE GRATING SPECTROGRAPH ADJUSTMENT

The adjustments described will apply particularly to a Czerny–Turner mounting, although most of them are generally useful.   The slit holder should be adjustable in height, and have provisions for leveling.   It should be mounted on a slide for motion along a line joining the slit and collimating mirror.   Both mirror mounts should allow for height adjustment, coarse rotation about a vertical axis, and fine adjustment of rotation about the vertical axis and about a horizontal axis tangent to the surface of the mirror. The grating table should have height and leveling adjustments, the latter with precision control if possible.   It must rotate smoothly about the vertical axis with a fast motion, as well as precision slow motion for setting exact angles.   The lock on this motion should not influence the setting. The plate holder must be adjustable in height and be mounted on a slide for motion along a line joining it and the camera mirror.   It should accept both a cassette for a photographic plate and a photomultiplier.   Since this spectrograph mounting is stigmatic, the slit should have a slide with cut outs for illuminating different sections of the slit, and the plateholder should have a provision for racking the cassette up and down.   No occulter is necessary.   The grating mounting on the turntable must allow small and carefully controllable rotations about three mutually perpendicular axes.

First, set the slit, mirrors, grating, and plateholder in approximately the correct positions.   To minimize aberrations the mirrors should be as close together as possible, and the slit and plateholder as close together as possible.   Some asymmetry may be desirable in the grating position between the slit and plateholder, as mentioned in one of the references.   Next set them all at the same height, using a transit.   Then point the slit base directly at the center of the collimating mirror, using the following procedure. Set up a mercury arc in front of the slit, but some distance away from it. Adjust its position so that light passing through the slit falls on the center of the collimating mirror. Adjust the mirror, grating and camera mirror so that the zeroth order of the grating falls approximately at the center of the plateholder. Then use a condensing lens to focus an image of the arc on the slit; light should now cover the entire face of the grating. At the plateholder, observe the light with a magnifier, while moving the slit assembly toward and away from the collimating mirror. Adjust the direction of the slit base until the image does not move sideways as the slit is moved. Of course the light will not remain in focus, but the center of the image should stay in the same place on the track.

The slit should next be made vertical. With the light source as above, place a lens behind the slit (on the mirror side) so that it forms a greatly enlarged image of the slit on a white card placed just in front of the collimating mirror. Place a plumb bob in front of the card, and adjust its position until its shadow bisects the image of the slit. Any deviation of the slit from vertical should be corrected.

The next step is to put the slit at the focal point of the collimating mirror. Remove the imaging lens from between the grating and slit. Rotate the grating until light in the zero order is returned to the center of the collimating mirror, and back out through the slit. Place a pinhole at the slit. (The pinhole should be 0.05 to 0.1 mm diameter, and can be formed by pushing a needle through a piece of brass shim stock.) Rotate the grating slightly until the image of the pinhole is slightly above the pinhole itself, on the backside of the material in which the pinhole is formed. Move the slit back and forth on its base until the image is in good focus. The slit is now in its proper position. It should be mentioned that this procedure is not valid if the grating blank is not flat. In this event, a plane mirror the size of the projected image of the grating when in use must be substituted for the grating. The mirror must be flat to one-eighth wavelength for best results.

A laser having a uniphase wavefront is needed for the next adjustments, although with care a strong but small light source can be used, such as a high pressure mercury arc. Set the laser outside the spectrograph, and point it through the center of the slit toward the center of the collimating mirror. Adjust the optical components until the beam strikes the center of the grating, center of the camera mirror at the zero order of the grating, and center of the plateholder as exactly as possible. Check the leveling of the grating turntable, and readjust the tilt of the grating face if necessary. Turn the grating first to one side and then the other; if the spectrum orders are high on one side and low on the other as observed at the plateholder, tilt the grating grooves in the appropriate direction.

This adjustment is most sensitive at the highest order on each side. Check the zero order, and repeat if necessary. If the zero order is diffracted to the center of the plateholder, but the orders on each side are both high or both low, then the axis of the grating turntable is not vertical, and must be adjusted. It may be necessary to spend considerable time with these adjustments.

To point the plateholder directly at the camera mirror, set up the instrument with a strong visible spectrum line at the center. Move the plateholder in and out and adjust the direction of the slide until the line does not move sideways. To focus the plate, take successive spectrograms at different distances from the camera mirror, and adjust position and tilt in accordance with the best focus over the plate. These settings will be constant for all angles of incidence if the grating is truly plane. The final step is correct placement of the optical bench outside the spectrograph. Since the laser points along the optical axis, it is an easy task to place a condensing lens on an optical bench, and adjust both ends of the bench so that the laser beam goes through the center of the lens, both horizontally and vertically.

As a final check, set up a mercury lamp with condensing lens, and adjust the grating so as to place the green line at the center of the plate. Put your eye right at the plate, and look at the camera mirror. It should be possible to see the entire face of the grating illuminated in green light, and the grating should appear to be in the center of the camera mirror. If light from the slit strikes the camera mirror directly (as well as the collimating mirror, of course), there will appear to be a bright slit image in the camera mirror. Block this light by placing a piece of black paper a few centimeters from the slit, at the appropriate position. Then rotate the grating to place the green line first at one side and then at the other of the plateholder, and make the same observations. The grating face should still be completely visible and completely filled with light, but off-center in the mirror.

A calibration curve for the wavelength as a function of grating angle can be made, for convenience in use of the instrument. The most useful plot is one of order-angstroms vs grating angle. For this purpose, set up a white light source, such as a xenon arc. Place a small wavelength measuring spectroscope at the center of the plateholder, pointing toward the camera mirror. At each setting of the grating, one or more orders of the spectrum can be seen in the spectroscope as sharp spectral lines, the number depending on the groove spacing. Record as many wavelengths as possible at each grating angle, multiply each by its corresponding order, and average; then plot the desired curve. The reproducibility of the settings should be good to 5 angstroms at worst.

It is sometimes difficult to be sure that a predisperser is allowing the correct wavelength into the instrument, particularly with a coarsely ruled grating.   If the desired wavelength is in the visible region of the spectrum it is easy to look at the light reaching the plateholder, and adjust the predisperser accordingly.   If the wavelength is in the ultraviolet or infrared, set the grating at the proper angle according to the calibration curve, adjust the predisperser to a corresponding wavelength in the visible region of the spectrum (in a different order from the desired wavelength), and substitute a photomultiplier for the eye at the plateholder.   Observe the successive maxima in photomultiplier output as the predisperser wavelength is gradually changed, until the desired order and wavelength are reached.   For example, suppose that 3641 Å is to be observed in the 15th order.   Other wavelengths which can appear at the same grating angle are 3901, 4201, 4551, 4965, and 5461 Å.   Set the predisperser visually so that it passes 5461 Å, place a photomultiplier at the plateholder, and observe its output while slowly changing the wavelength passed into the spectrograph by the predisperser.   Maxima in the output will be observed at each of the above wavelengths.   When the fifth maximum is reached, the predisperser is properly set at 3641 Å.   A white-light source can be used in making this setting, if the source to be observed is not bright enough.

Bibliography

The list of references given here is not intended to be exhaustive, but it is complete enough to provide a guide to the literature available. Very few references prior to 1940 are included, since they are mainly of historical interest. Comments have been appended to a few titles.

CHAPTER 1

FOWLES, G. R., *Introduction to Modern Optics*, Holt, Rinehart, and Winston, New York (1968).

JENKINS, F. A., and H. E. WHITE, *Fundamentals of Optics*, McGraw-Hill, New York (1957).

HARRISON, G. R., R. C. LORD, and J. R. LOOFBOUROW, *Practical Spectroscopy*, McGraw-Hill, New York (1948).

SAWYER, R. A., *Experimental Spectroscopy*, Dover, New York (1963).

STONE, J. M., *Radiation and Optics*, McGraw-Hill, New York (1963).

BORN, M., and E. WOLF, *Principles of Optics*, MacMillan, New York (1964).

CHAPTER 2

Grating Ruling and Production. References are grouped according to the various laboratories.

MICHELSON, A. A., "Ruling and Testing of a Ten-inch Diffraction Grating," *Am. Phil. Soc. Proc.* **54**, 137 (1915).

WOOD, R. W., "Improved Diffraction Gratings and Replicas," *J. Opt. Soc. Am.* **34**, 509 (1944).

STRONG, J., "New Johns Hopkins Ruling Engine," *J. Opt. Soc. Am.* **41**, 3 (1951).

————, "The Johns Hopkins University Diffraction Gratings," *J. Opt. Soc. Am.* **50**, 1148 (1960).

MERTON, T., "On the Reproduction and Ruling of Diffraction Gratings," *Proc. Roy. Soc.* **A201**, 187 (1949).

DEW, G. D., and L. A. SAYCE, "On the Production of Diffraction Gratings. I. The Copying of Plane Gratings," *Proc. Roy. Soc.* **A207**, 278 (1951).

HALL, R. G. N., and L. A. SAYCE, "On the Production of Diffraction Gratings. II. The Generation of Helical Rulings and the Preparation of Plane Gratings therefrom," *Proc. Roy. Soc.* **A215**, 525 (1952).

DEW, G. D., "On the Preservation of Groove Form in Replicas of Diffraction Gratings," *J. Sci. Instrum.* **29**, 277 (1952).

————, "On the Preparation of Plane Diffraction Grating Replicas from Helical Rulings," *J. Sci. Instrum.* **30**, 229 (1953).

————, "On Preparing Plastic Copies of Diffraction Gratings—An Extension to the Merton—N.P.L. Process," *J. Sci. Instrum.* **33**, 348 (1956).

BABCOCK, H. D., and H. W. BABCOCK, "The Ruling of Diffraction Gratings at the Mount Wilson Observatory," *J. Opt. Soc. Am.* **41**, 776 (1951).

BABCOCK, H. W., "Control of a Ruling Engine by a Modulated Interferometer," *Appl. Opt.* **4**, 415 (1962).

HARRISON, G. R., "The Production of Diffraction Gratings. I. Development of the Ruling Art," *J. Opt. Soc. Am.* **39**, 413 (1949).

HARRISON, G. R., and J. E. ARCHER, "Interferometric Calibration of Precision Screws and Control of Ruling Engines," *J. Opt. Soc. Am.* **41**, 495 (1951).

HARRISON, G. R., and W. H. CULVER, "Ruling of Test Gratings with Interferometric Control," *J. Opt. Soc. Am.* **41**, 870 (1951).

HARRISON, G. R., and G. W. STROKE, "Interferometric Control of Grating Ruling with Continuous Carriage Advance," *J. Opt. Soc. Am.* **45**, 112 (1955).

HARRISON, G. R., N. STURGIS, S. C. BAKER, and G. W. STROKE, "Ruling of Large Diffraction Gratings with Interferometric Control," *J. Opt. Soc. Am.* **47**, 15 (1957).

HARRISON, G. R., N. STURGIS, S. P. DAVIS, and Y. YAMADA, "Interferometrically Controlled Ruling of Ten-inch Diffraction Gratings," *J. Opt. Soc. Am.* **49**, 205 (1959).

HARRISON, G. R., "Interferometric Control in the Ruling of Large Diffraction Gratings," *Sci. of Light* **6**, 26 (1967).

STROKE, G. W., "Diffraction Gratings," *Handbuch der Physik* **29**, 426 (1967).

GERASIMOV, F. M., I. A. TEL'TEVSKII, S. S. NAUMOV, S. N. SPIZHARSKII, and S. V. NESMELOV, "Diffraction Gratings from the State Optical Institute," *Opt. i Spekt.* **4**, 779 (1958).

RASSUDOVA, G. N., and F. M. GERASIMOV, "Precision Diffraction Gratings for Metrology," *Opt. and Spectr.* **11**, 136 (1961).

FUJIOKA, Y., and Y. SAKAYANAGI, "Endeavour on Ruling Grating in Japan," *Sci. of Light* **2**, 1 (1952).

SAKAYANAGI, Y., "Ruling of a Curved Grating," *Sci. of Light* **3**, 79 (1955).

LEIBHARDT, E., and J. DU BOIS, "The Interferometric Control System of the Diffraction Products' Ruling Engine," *Dev. in Appl. Spectr.* **4**, 495 (1965).

HORSFIELD, W. R., "Ruling Engine with Hydraulic Drive," *Appl. Opt.* **4**, 189 (1965).

DAVIES, D. A., and G. M. STIFF, "Diffraction Grating Ruling in Australia," *Appl. Opt.* **8**, 1379 (1969).

Theory and Practice of Testing Diffraction Gratings.

General Testing

HARRISON, G. R., "The Testing and Use of Concave Diffraction Gratings," *Proc. Seventh Spectroscopy Conf.*, Technology Press, Cambridge, 1940.

MARÉCHAL, A., "The Control of Grating Quality by Phase Contrast" (*Contraste de Phase et Contraste par Interference*), *Revue d'Optique* **204** (1952).

INGELSTAM, E., and E. DJURLE, "The Study of Diffraction Grating Characteristics by Simplified Phase Contrast Methods," *J. Opt. Soc. Am.* **43**, 572 (1953).

STROKE, G. W., "Interferometric Measurement of Wave-front Aberrations in Gratings and Echelles," *J. Opt. Soc. Am.* **45**, 30 (1955).

POGGIO, M. A., "Interpretation of the Fringe Structure of a Wave Front Reflected from a Concave Diffraction Grating," *Rev. Univ. Nac. La Plata* **3**, 51 (1961).

SHIMIZU, K., "On the Ruling Errors of a Diffraction Grating and Their Effects on the Spectral Lines," *Japan J. Appl. Phys.* **4**, 555 (1965).

BIRCH, K. G., "Interferometric Examination of the Ruling Errors of a Concave Grating," *J. Sci. Instrum.* **43**, 243 (1966).

Diffraction Anomalies

WOOD, R. W., "Anomalous Diffraction Gratings," *Phys. Rev.* **48**, 928 (1935).

FANO, U., "Theory of Anomalous Diffraction Gratings and of Quasi-Stationary Waves on Metallic Surfaces (Sommerfeld's Waves)," *J. Opt. Soc. Am.* **31**, 213 (1941).

PALMER, C. H., "Parallel Diffraction Grating Anomalies," *J. Opt. Soc. Am.* **42**, 269 (1952).

TWERSKY, V., "On a Multiple Scattering Theory of the Finite Grating and the Wood Anomalies," *J. Appl. Phys.* **23**, 1099 (1952).

TWERSKY, V., "Remarks on the Theory of Grating Anomalies," *J. Opt. Soc. Am.* **42**, 855 (1952).

PALMER, C. H., JR., "Diffraction Grating Anomalies, II. Coarse Gratings," *J. Opt. Soc. Am.* **46**, 50 (1956).

STEWART, J. E., and W. S. GALLAWAY, "Diffraction Anomalies in Grating Spectrophotometers," *Appl. Optics* **1**, 421 (1962).

HESSEL, A., and A. A. OLIVER, "A New Theory of Wood's Anomalies on Optical Gratings," *Appl. Opt.* **4**, 1275 (1965).

Resolving Power

STRONG, J., "Resolving Power Limitations of Grating and Prism Spectrometers," *J. Opt. Soc. Am.* **39**, 320 (1949).

HULTHÉN, E., and U. UHLER, "Examination of the Resolving Power of the Diffraction Grating," *Ark. Fys.* **3**, 393 (1951).

RANK, D. H., "Theoretical Resolving Power of Diffraction Gratings," *J. Opt. Soc. Am.* **42**, 279 (1952).

VOLLMER, R., "Dispersion and Resolution of a Plane Grating Under High Angles of Incidence and Diffraction," *Optik* **10**, 497 (1953).

SHARMA, P., and M. S. SODHA, "On Dependence of Resolving Power of Prism, Grating and Reflecting Echelon on Stage of Resolution and Detecting Instrument," *Indian J. Phys.* **28**, 437 (1954).

RANK, D. H., J. N. SHEARER, and J. M. BENNETT, "Quantitative Method for Measuring the Resolution of a Large Grating," *J. Opt. Soc. Am.* **45**, 762 (1955).

CHATURVEDI, K. C., "Resolution of Spectral Lines of Unequal Intensity in Grating, Reflecting Echelon and Prism," *Optik* **17**, 70 (1960).

HARRISON, G. R., and G. W. STROKE, "Attainment of High Resolution with Diffraction Gratings and Echelles," *J. Opt. Soc. Am.* **50**, 1153 (1960).

STROKE, G. W., "Attainment of High Resolution Gratings by Ruling under Interferometric Control," *J. Opt. Soc. Am.* **51**, 1321 (1961).

Ghosts and Satellites

LYMAN, T., "False Spectra with Rowland Grating," *Phys. Rev.* **12**, 1 (1901).

WOOD, R. W., "An Experimental Study of Grating Errors and 'Ghosts'," *Phil. Mag.* **48**, 497 (1924).

GALE, H. G., "Rowland Ghosts," *Astrophys. Jour.* **85**, 49 (1937). Excellent photographs of Rowland Ghosts.

POMERANCE, H. S., and H. G. BEUTLER, "Identification of Orders and Ghosts in Grating Spectra by Diffracting Slits," *Rev. Mod. Phys.* **14**, 66 (1942).

FINKELSTEIN, N. A., C. H. BRUMLEY, and R. J. MELTZER, "The Reduction of Ghosts in Diffraction Grating Spectra," *J. Opt. Soc. Am.* **42**, 121 (1952).

SAGE, S. H., and R. E. SWING, "Diffraction Grating Study: Derivation and Experimental Verification of the Cause of Tilted Spectral Images," *J. Opt. Soc. Am.* **45**, 256 (1955).

DJURLE, E., "On the Origin of Unsymmetrical Rowland Ghosts in Optical Gratings," *Ark. Physik* **8**, 383 (1954).

PIERCE, A. K., "Performance of an 8-inch Babcock Grating in a Large Vacuum Spectrograph," *J. Opt. Soc. Am.* **47**, 6 (1957).

SHENSTONE, A. G., "New Vagary of a Concave Grating," *J. Opt. Soc. Am.* **53**, 1253 (1963).

Blaze, and Grating Efficiency

STAMM, R. F., and J. J. WHALEN, "Energy Distribution of Diffraction Gratings as a Function of Groove Form." (Calculations by an Equation of H. A. Rowland), *J. Opt. Soc. Am.* **36**, 2 (1946).

FRIEDL, W., and B. HARTENSTEIN, "Energy Distribution of Diffraction Gratings as a Function of Groove Form," *J. Opt. Soc. Am.* **45**, 398 (1955).

SEYA, M., and K. GOTO, "On the Energy Distributions of Diffracted Light from Plane Gratings," *Sci. of Light* **5**, 119 (1956).

HARTENSTEIN, B., and W. KRIEDL, "The Intensity Distribution from a Plane Grating with a Triangular Groove Profile," *Optik* **14**, 119 (1957).

SAKAYANAGI, Y., "An Example of Making a Bright Grating and Its Application," *Sci. of Light* **5**, 110 (1956).

MARÉCHAL, A., and G. W. STROKE, "The Origin of the Effects of Polarization and of Diffraction in Optical Gratings," *C. R. Acad. Sci.* **249**, 2042 (1959).

SCHAFFER, O., "On the Influence of the Form of the Rulings of a Plane Phase Grating on the Light Distribution of the Fraunhofer Diffraction Pattern," *Optik* **16**, 200 (1959).

MATTIG, W., and E. H. SCHROTLR, "Test Results on a High-Intensity Blazed Diffraction Grating," *Optik* **16**, 339 (1959).

STROKE, G. W., "Theoretical and Experimental Study of Two Aspects by Diffraction by Gratings: (I) Causes of Defects in Spectrum Line Profiles and (II) Electromagnetic Theory of Light Distribution among Different Orders," *Rev. Opt.* **39**, 291 (1960).

YAKOVLEV, E. A., and F. M. GERASIMOV, "Experimental Study of the Intensity Distribution in the Spectrum of Diffraction Gratings for Polarized Light," *Opt. and Spectr.* **10**, 50 (1961).

PALMER, C. H., JR., "Comment on the Article by Yakovlev and Gerasimov on the Intensity Distribution in the Spectrum of Diffraction Gratings," *J. Opt. Soc. Am.* **51**, 1438 (1961).

YAKOVLEV, E. A., and F. M. GERASIMOV, "Comments on a Note by C. H. Palmer," *Opt. and Spectr.* **13**, 58 (1962).

HIDALGO, A., J. PASTOR, and J. M. SERRATOSA, "Effect of Polarization in the Distribution of Energy Diffracted by Gratings," *J. Opt. Soc. Am.* **52**, 1081 (1962).

STROKE, G. W., "Attainment of High Efficiencies in Blazed Optical Gratings by Avoiding Polarization in the Diffracted Light," *Phys. Letters* **5**, 45 (1963).

NAMIOKA, T., and M. SEYA, "The Efficiency of Concave Grating," *Sci. of Light* **15**, 1 (1966).

YAKOVLEV, E. A., "Calculation of the Distribution of Intensities by a Diffraction Grating in Polarized Light," *Opt. and Spectr.* **19**, 233 (1965).

WATANABE, K., "Improvement of Grating Efficiency in the Vacuum Ultraviolet by Platinizing," *J. Opt. Soc. Am.* **43**, 318 (1953).

SPRAGUE, G., D. H. TOMBOULLION, and D. E. BEDO, "Calculations of Grating Efficiency in the Soft X-Ray Region," *J. Opt. Soc. Am.* **45**, 756 (1955).

HASS, G. H., W. R. HUNTER, and R. TOUSEY, "Influence of Purity, Substrate Temperature and Aging Conditions on the Extreme Ultraviolet Reflectance of Evaporated Aluminum," *J. Opt. Soc. Am.* **47**, 1070 (1957).

CRISP, R. S., "The Reflectivity of Glass and Aluminum Gratings at Grazing Incidence Below 1600 Å," *Optica Acta* **8**, 137 (1961).

WILKINSON, P. G., and D. W. ANGEL, "Evaluation of a Specially Coated Concave Diffraction Grating in the Vacuum Ultraviolet," *J. Opt. Soc. Am.* **52**, 1120 (1962).

REEVES, A. M., and W. H. PARKINSON, "Efficiencies of Gold and Platinum Gratings in the Vacuum Ultraviolet," *J. Opt. Soc. Am.* **53**, 41 (1963).

TOMBOULIAN, D. H., and W. E. BEHRING, "A Method for Comparing the Reflecting Power of Concave Gratings in the Soft X-ray Region," *Appl. Optics* **3**, 501 (1964).

MORSE, A. L., and G. L. WEISSLER, "The Efficiency of Concave Gratings in the Extreme Ultraviolet," *Sci. of Light* **15**, 22 (1966).

HANSON, W. F., and E. T. ARAKAWA, "Reflectances of Concave Diffraction Gratings for Polarized Vacuum Ultraviolet," *J. Opt. Soc. Am.* **56**, 124 (1966).

Stray Light and Scattered Light

INGELSTAM, E., and E. DJURLE, "A Method of Determining Common Stray Light for Ruled Gratings and Other Spectral Apparatus," *Ark. Fys.* **3**, 63 (1951).

VON ALVENSLEBEN, A., "The Apparatus Functions, Grating Ghosts and Stray Light for the Two New Plane Gratings in the Göttingen Solar Spectrograph," *Z. Astrophys.* **43**, 63 (1957).

MARÉCHAL, A., "The Residual Scattering of Polished Surfaces and Gratings," *Optica Acta* **5**, 70 (1958).

MILLAR, R. F., "Scattering by a Grating," *Am. J. Phys.* **39**, 81, 104 (1961).

CHAPTERS 3 AND 4

Theory of Grating Spectrometers, including Aberrations

ROWLAND, H. A., "Gratings in Theory and Practice. I." *Ast. and Astro Phys.* **12**, 129 (1893).

BEUTLER, H. G., "The Theory of the Concave Grating," *J. Opt. Soc. Am.* **35**, 311 (1945).

TONO, U., "On the Theory of Imperfect Diffraction Gratings," *J. Opt. Soc. Am.* **38**, 921 (1948).

MENZELAND, E., and C. MENZEL, "Diffraction Effects with Optical Gratings According to Intensity and Phase," *Optik* **3**, 247 (1948).

————, "On the Imagery of Optical Gratings," *Optik* **4**, 22 (1948).

VON KEUSSLER, V., "On the Geometrical Conditions for Refraction through a Prism and Diffraction through a Grating," *Z. Astrophys.* **24**, 247 (1948).

TORALDO DI FRANCIA, G., "Third-order Aberrations of the Diffraction Grating," *Nuovo Cimento* **6**, 24 (1949).

FASTIE, W. G., "Image Forming Properties of the Ebert Monochromator," *J. Opt. Soc. Am.* **42**, 647 (1952).

NAMIOKA, T., "Grating Theories and their Application to Grating Instruments for Vacuum Ultraviolet Spectroscopy," *J. Quant. Spectr. Rad. Tran.* **2**, 697 (1952).

LIPPMANN, B. A., "Note on the Theory of Gratings," *J. Opt. Soc. Am.* **43**, 408 (1953).

GREINER, H., and E. SCHAFFER, "Seya's Theory of the Concave Grating Spectrometer," *Optik* **14**, 263 (1957).

NAMIOKA, T., "Theory of the Concave Grating. I." *J. Opt. Soc. Am.* **49**, 446 (1959).

————, "Theory of the Concave Grating. II. Application of the Theory to the Off-plane Eagle Mounting in a Vacuum Spectrograph," *J. Opt. Soc. Am.* **49**, 460 (1959).

————, "Theory of the Concave Grating. III. Seya-Namioka Monochromator," *J. Opt. Soc. Am.* **49**, 951 (1959).

MIELENZ, K. D., "Theory of Grating Spectroscopic Instruments with Rectangular Apertures," *Optik* **16**, 458 (1959).

KOZLENKOV, A. I., "An Investigation of Aberrations of a Concave Grating," *Opt. and Spectr.* **8**, 365 (1960).

GREINER, H., and E. SCHAFFER, "Theory of a Concave Grating Spectrometer," *Optik* **15**, 51 (1958).

SHELKOVA, O. P., "A Simplified Method of Designing Monochromators with Plane Diffraction Gratings," *Svetotekhnika* **5**, 14 (1961).

ROSENDAHL, G. R., "Contributions to the Optics of Mirror Systems and Gratings with Oblique Incidence. I. Ray Tracing Formulas for the Meridional Plane," *J. Opt. Soc. Am.* **51**, 1 (1961).

————, "Contributions to the Optics of Mirror Systems and Gratings with Oblique Incidence. II. A Discussion of Aberrations," *J. Opt. Soc. Am.* **52**, 408 (1962).

————, III. Some Applications, *J. Opt. Soc. Am.* **52**, 412 (1962).

MIYAKI, K. P., and T. KATAYAMA, "On the Mounting of Concave Grating Suitable for Photoelectric Spectrometer Use. Discussion on Results Obtained in Theoretical Treatise," *Sci. of Light* **11**, 1 (1962).

————, ————, "Discussion of the Use of the Eagle Mounting," *Sci. of Light* **11**, 10 (1962).

————, "Generalized Seya Mounting," *Sci. of Light* **11**, 21 (1962).

SAGAWA, T., "The Aberrations of the Concave Grating at Grazing Incidence," *Sci. Rep. Tohoku* **46**, 119 (1962).

MURTY, M. V., "Use of Convergent and Divergent Illumination with Plane Gratings," *J. Opt. Soc. Am.* **52**, 768 (1962).

WELFORD, W. T., "Tracing Skew Rays through Concave Diffraction Gratings," *Optica Acta* **9**, 389 (1962).

JAEGLÉ, P., "Calculation of the Useful Width of a System of Two Concave Gratings at Grazing Incidence," *J. Phys.* **24**, 179 (1963).

KASTNER, S. O., and W. M. NEUPERT, "Image Construction for Concave Gratings at Grazing Incidence, by Ray Tracing," *J. Opt. Soc. Am.* **53**, 1180 (1963).

SHAFER, A. B., L. R. MEGILL, and L. DROPPLEMAN, "Optimization of the Czerny–Turner Spectrometer," *J. Opt. Soc. Am.* **54**, 879 (1964).

MIELENZ, K. D., "Theory of Mirror Spectrographs. I. Astigmatic Illumination of Plane Gratings and Prisms. II. General Theory of Focal Surfaces and Slit Curvatures. III. Focal Surfaces and Slit Curvature of Ebert and Ebert–Fastie Spectrographs," *J. Res. Nat. Bur. Stand.* **68C**, 195 (1964).

WILLSTROP, R. V., "A Wide-Field Coma-Free All-Reflection Plane Grating Spectrometer," *Monthly Not. Roy. Astron. Soc.* **130**, 233 (1965).

ROUSE, JR., P. E., B. BRIXNER, and J. V. KLINE, "Optimization of a 4-m Asymmetric Czerny–Turner Spectrograph," *J. Opt. Soc. Am.* **59**, 955 (1969).

READER, J., "Optimizing Czerny–Turner Spectrographs: A Comparison between Analytic Theory and Ray Tracing," *J. Opt. Soc. Am.* **59**, 1189 (1969).

Unconventional Diffraction Gratings and Their Mountings

HABER, H., "The Torus Grating," *J. Opt. Soc. Am.* **40**, 153 (1950).

MERTON, T. R., "Use of Diffraction Gratings Ruled on Cylinders," *Nature* **166**, 866 (1950).

SAKAYANAGI, Y., "Theory of Grating with Circular Grooves," *Sci. of Light* **3**, 1 (1954).

DYSON, J., "Circular and Spiral Diffraction Gratings," *Proc. Roy Soc.* **A248**, 93 (1958).

SCHLEPTKIN, Y. P., "Aspherical Diffraction Grating with One Plane of Symmetry. I. Aberrations of an Aspherical Grating." PB141047T-4, Office of Technical Services, Washington (1958).

———, "An Aspherical Diffraction Grating with One Plane of Symmetry. II. Permissible Values of Aberrations. The Range of Application and Efficiency of an Aspherical Grating." PB141047T-9, Office of Technical Services, Washington (1958).

GREINER, H., and E. SCHAFFER, "Astigmatism of the Concave Grating with Spherical or Toroidal Surfaces," *Optik* **16**, 288 (1959).

———, "Torus Grating Spectrometer," *Optik* **16**, 350 (1959).

MURTY, M. V., "Spherical Zone-plate Diffraction Grating," *J. Opt. Soc. Am.* **50**, 923 (1960).

NAMIOKA, T., "Theory of the Ellipsoidal Concave Grating. I." *J. Opt. Soc. Am.* **51**, 4 (1961).

———, "Theory of the Ellipsoidal Concave Grating. II. Application of the Theory to the Specific Grating Mountings," *J. Opt. Soc. Am.* **51**, 13 (1961).

GALE, B., "The Theory of Variable Spacing Gratings," *Opt. Acta* **13**, 41 (1966).

SAKAYANAGI, Y., "A Stigmatic Concave Grating with Varying Spacing," *Sci. of Light* **16**, 129 (1967).

Diffraction Grating Mountings

WADSWORTH, F. L. O., "Fixed Aim Concave Grating Spectroscopes," *Astr. J.* **2**, 370 (1895).

EAGLE, A., "On a New Mounting for a Concave Grating," *Ast. J.* **31**, 120 (1910).

KING, A. S., "A Vertical Adaptation of the Rowland Mounting for a Concave Grating," *Ast. J.* **40**, 205 (1914).

BREWINGTON, G. P., "An Improvement in the Design of the Concave-grating Spectrograph," *Rev. Sci. Instrum.* **13**, 501 (1942).

JARRELL, R. F., "A Stigmatic Grating Spectrograph for Industrial Laboratories," *J. Opt. Soc. Am.* **32**, 666 (1942).

ROIG, J., L. BOURDELET, and J. ROUSSEAU, "Construction of a Stigmatic Mounting for a Concave Grating," *Rev. Opt.* **31**, 286 (1952).

KOCH, G. P., W. J. TAYLOR, and H. L. JOHNSTON, "A Recording Grating Spectrometer with Linear Wavelength Scale," *J. Opt. Soc. Am.* **41**, 125 (1951).

FINKELSTEIN, N. A., "An Efficient Grating Mount," *J. Opt. Soc. Am.* **41**, 179 (1951).

GILLIESON, A. H. C. P., "A New Spectrographic Diffraction Grating Mounting," *J. Sci. Instrum. Phys. Ind.* **26**, 335 (1949).

FASTIE, W. G., "A Small Plane Grating Monochromator," *J. Opt. Soc. Am.* **42**, 641 (1952).

STRAAT, H. W., "Compensation of Astigmatic Errors in a Grating Spectrograph," *J. Opt. Soc. Am.* **43**, 593 (1953).

SHENSTONE, A. G., "Wadsworth Mounting for Grating," *J. Opt. Soc. Am.* **43**, 706 (1953).

FISHER, R., "Resolution and Dispersion of a Concave Grating Spectrometer Using a Photon Multiplier Detector," *J. Opt. Soc. Am.* **44**, 665 (1954).

NAMIOKA, T., "Construction of a Grating Spectrometer," *Science of Light* **3**, 15 (1954).

LEE, T., "The Design and Construction of a 22-foot Direct-Reading Optical Spectrometer," *Appl. Spectroscopy* **8**, 174 (1954).

WEINARD, J., "The Construction of a Combination Grating Spectrograph of High Dispersion," *Z. Angew. Phys.* **7**, 584 (1955).

JARRELL, R. F., "Stigmatic Plane Grating Spectrograph with Order Sorter," *J. Opt. Soc. Am.* **45**, 259 (1955).

GOODY, R. M., "A Simple Grating Spectrometer for Sky-emission Studies," *Quart. J. Roy. Meteorol. Soc.* **83**, 517 (1957).

KING, G. W., "A 20-foot Ebert Grating Spectrograph," *J. Sci. Instrum.* **35**, 11 (1958).

GREINER, H., and E. SCHAFFER, "Contribution to the Practical Attainment of a Spectrometer with Rotatable Concave Grating," *Z. Instrum. Kde* **66**, 188 (1958).

SREEKANTATH, G. M., and C. A. VERGHESE, "On the Optical Properties of a Biprism Combined with a Coarse Grating," *J. Sci. Instrum.* **35**, 150 (1958).

WELFORD, W. T., "Stigmatic Ebert-type Plane Grating Mounting," *J. Opt. Soc. Am.* **53**, 766 (1963).

LILLER, W., "Concave Gratings for Astronomical Spectrographs and Spectrometers," *Appl. Optics* **2**, 187 (1963).

BRIL, A., and W. VAN MEURS-HOEKSTRA, "Use of Diffraction Gratings in Small Mirror Monochromators," *Z. Instrum. Kde* **71**, 232 (1963).

LANDON, D. O., "A New Grazing Incidence Spectrometric Mounting," **3, 115** (1964).

BAKER, S. C., "Stigmatic Eberts-type Plane Grating Mounting," *J. Opt. Soc. Am.* **154,** 271 (1964).

KUDO, K. "Plane Grating Monochromators of Ebert, Pfund, and Czerny-Turner Types," *Sci. of Light* **9,** 1 (1960).

————, "Plane Grating Monochromator of Littrow Type," *Sci. of Light* **9,** 65 (1960).

HILL, R. A., and E. H. BECKNER, "A Rapid Scan Spectrograph for Plasma Spectroscopy," *Appl. Opt.* **3,** 929 (1964).

SERGENT-ROZEY, M., "Method of Improving the Image Formation in a Grating Spectrometer," *Rev. Opt.* **44,** 193 (1965).

Diffraction Grating Mountings Especially for the Vacuum Ultraviolet.

SEYA, M., "A New Mounting of Concave Grating Suitable for a Spectrometer," *Sci. of Light* **2,** 8 (1952).

NAMIOKA, T., "Construction of a Grating Spectrometer," *Sci. of Light* **3,** 15 (1954).

LUSCHER, E., "A New Grating Spectrograph for the Schumann Region with Photoelectric Recording," *Helv. Phys. Acta* **28,** 492 (1955).

ROBIN, S., and ST. ROBIN, "Vacuum Spectrograph and Concave Grating in Normal Incidence for the Far Ultraviolet," *J. Phys. Radium* **17,** 976 (1956).

OAKA, R., "Grating Mounting for Vacuum Ultraviolet Monochromator," *Sci. of Light* (*Tokyo*) **7,** 23 (1958).

NAMIOKA, T., "Design of High-resolution Monochromator for the Vacuum Ultraviolet. An Application of Off-plane Eagle Mounting," *J. Opt. Soc. Am.* **49,** 961 (1959).

RENSE, W. A., and T. VIOLETT, "Method of Increasing the Speed of a Grazing-Incidence Spectrograph," *J. Opt. Soc. Am.* **49,** 139 (1959).

LUKIRSKU, A. P., and E. P. SAVENOV, "Use of Diffraction Grating and Echelettes in the Ultrasoft X-Ray Region," *Opt. and Spectr.* **14,** 147 (1963).

DOUGLAS, A. E., and J. G. POTTER, "A Ten-Meter Grating Spectrograph for the Vacuum Ultraviolet," *Appl. Opt.* **1,** 727 (1962).

LANDON, D. O., "A New Grazing Incidence Spectrometric Mounting," *Appl. Opt.* **3,** 115 (1964).

Diffraction Grating Mountings Especially for the Infrared.

OETJEN, R. A., W. H. HAYNIE, W. M. WARD, R. L. HANSLER, H. E. SCHAUWECKER, and E. E. BELL, "An Infrared Spectrograph for Use in the 40–159-Micron Spectral Region," *J. Opt. Soc. Am.* **42,** 559 (1952).

GOULDEN, J. D. S., "The Use of Diffraction Gratings with an Infrared Spectrometer," *J. Sci. Instrum.* **29,** 215 (1952).

HADNI, A., "Grating Spectrometer for the Far Infrared," *Rev. Opt.* **33,** 576 (1954).

AMOT, G., "Grating Spectrographs in Use at the P. C. B. Infrared Laboratory," *Rev. Opt.* **33**, 642 (1954).

SAKAYANAGI, Y., and Y. SATO, "A Near Infrared Spectrometer Using a Plane Grating," *Science of Light* **3**, 84 (1955).

PLYLER, E. K., and N. ACQUISTA, "Small Grating Spectrometer for the Far Infrared Region," *J. Chem. Phys.* **23**, 752 (1955).

LORD, R. C., and T. K. MCCUBBIN, JR., "High-Resolution Spectroscopy in the 3-5 Micron Region with a Small Grating Spectrometer," *J. Opt. Soc. Am.* **45**, 441 (1955).

HADNI, A., "Small Grating Spectrometer for the Far Infrared," *J. Phys. Radium* **17**, 77 (1956).

LORD, R. C., and T. K. MCCUBBIN, JR., "Infrared Spectroscopy from 5 to 200 Microns with a Small Grating Spectrometer," *J. Opt. Soc. Am.* **47**, 689 (1957).

WHITE, J. U., N. L. ALPERT, A. G. DE BELL, and F. M. CHAPMAN, "Infrared Grating Spectrophotometer," *J. Opt. Soc. Am.* **47**, 358 (1957).

FORD, M. A., W. C. PRICE, and G. R. WILKINSON, "A High-resolution Grating Spectrometer for the Infrared Region," *J. Sci. Instrum.* **35**, 55 (1958).

YOSHINAGA, H., S. FUJITA, S. MINAMI, A. MITSUISHI, R. A. OETJEN, and Y. YAMATA, "Far Infrared Spectrograph for Use from the Prism Spectral Region to about 1-mm Wavelength," *J. Opt. Soc. Am.* **48**, 315 (1958).

RAO, K. N., and H. H. NIELSEN, "Six-Meter Focal Length Ebert-Type Infrared Spectrograph. I. Details of the Spectrograph," *Appl. Opt.* **2**, 1123 (1963).

RAO, K. N., and P. E. FRALEY, "Six-Meter Focal Length Ebert-Type Infrared Spectrograph. II. Precision of Measurements." *Appl. Opt.* **2**, 1127 (1963).

TRIAILLY, E. A., C. D. COURTOY, A. MARTEGANI, and M. DE HEMPTINNE, "Grating Spectrometer of High Resolution for the Study of Infrared Spectra," *Ann. Soc. Sci. Bruxelles* **68**, 121 (1964).

RUSSELL, J. W., and H. L. STRAUSS, "Czerny–Turner Far Infrared Spectrometer for the 300-10 cm⁻¹ Region." *Appl. Opt.* **4**, 1131 (1965).

Diffraction Gratings in Immersion

ROGERS, G. L., "Total Internal Reflection and Huygen's Construction: The Immersion Grating," *Nature* **160**, 25 (1947).

HULTHÉN, E., and NEUHAUS, H., "Diffraction Gratings in Immersion," *Ark. f. Fysik* **8**, 343 (1954).

Luminosities of Spectrometers

JACQUINOT, P., "Luminosities Compared for Prism and Grating Spectrometers," *Rev. Opt.* **33**, 653 (1954).

————, "The Luminosity of Spectrometers with Prisms, Gratings, or Fabry-Perot Etalons," *J. Opt. Soc. Am.* **44**, 761 (1954).

————, "Luminosity of Spectrometers," *J. Opt. Soc. Am.* **45**, 996 (1955).

PERRY, J. W., "Luminosity of Spectrometers," *J. Opt. Soc. Am.* **45**, 995 (1955).

Multiple Diffraction from Gratings

JENKINS, F. A., and L. W. ALVAREZ, "Successive Diffractions by a Concave Grating," *J. Opt. Soc. Am.* **42**, 699 (1952).

RANK, D. H., and T. A. WIGGINS, "Double-passing a Plane Grating," *J. Opt. Soc. Am.* **42**, 983 (1952).

FASTIE, W. G., and W. M. SINTON, "Mutliple Diffraction in Grating Spectroscopy," *J. Opt. Soc. Am.* **44**, 103 (1954).

RANK, D. H., A. H. GUENTHER, C. R. BURNETT, and T. A. WIGGINS, "Examples of High Resolution Obtainable by Double Passing a High Quality Grating," *J. Opt. Soc. Am.* **47**, 631 (1957).

KARPINSKII, U. N., "Compensation of Rowland's 'Ghosts' in Spectrographs with Double Diffraction at the Grating," *Opt. i Spekt.* **8**, 207 (1960).

SHIMOMURA, T., "A Multi-Pass Spectrometer," *Japan J. Appl. Phys.* **3**, 459 (1964).

Echelle Spectrographs

WOOD, R. W., "The Use of Echellete Gratings in High Orders," *J. Opt. Soc. Am.* **37**, 733 (1947).

HARRISON, G. R., "The Production of Diffraction Gratings. II. The Design of Echelle Gratings and Spectrographs," *J. Opt. Soc. Am.* **39**, 522 (1949).

HARRISON, G. R., J. E. ARCHER, and J. CAMUS, "A Fixed-Focus Broad-Range Echelle Spectrograph of High Speed and Resolving Power," *J. Opt. Soc. Am.* **42**, 706 (1952).

HARRISON, G. R., S. P. DAVIS, and H. J. ROBERTSON, "Precision Measurement of Wavelengths with Echelle Spectrographs," *J. Opt. Soc. Am.* **43**, 853 (1953).

RANK, D. H., D. P. EASTMAN, W. B. BIRTLEY, G. SKORINKO, and T. A. WIGGINS, "Echelle-Type Spectrograph for the Near Infrared," *J. Opt. Soc. Am.* **50**, 821 (1960).

MAKINO, I., "On Double Diffraction Echelette Grating," *Sci. of Light* **13**, 1 (1964).

CHAPTER 5

MIELENZ, K. D., "Spectroscopic Slit Images in Partially Coherent Light," *J. Opt. Soc. Am.* **57**, 66 (1967).

Kodak Plates and Films for Science and Industry, Eastman Kodak Co., Rochester, N. Y.

POPE, T. P., and T. B. KIRBY, "Modified Hypersensitization Procedure for Eastman Kodak I–Z Spectroscopic Plates," *J. Opt. Soc. Am.* **57**, 951 (1967).

DOUGLAS, A. E., and G. HERZBERG, "Separation of Overlapping Orders of a Concave Grating Spectrograph in the Vacuum Ultraviolet Region," *J. Opt. Soc. Am.* **47**, 625 (1957).

RASSUDOVA, G. N., and F. M. GERASIMOV, "Diffraction Gratings for Separation of Spectral Orders," *Opt. i Spek.* **6**, 826 (1959).

READER, J., L. C. MARQUET, and S. P. DAVIS, "Predisperser for High-Resolution Grating Spectrographs," *Appl. Opt.* **2**, 963 (1963).

BLACKWELL, H. E., G. S. SHIPP, M. OGAWA, and G. L. WEISSLER, "Properties of a Plane Grating Predisperser used with a Grazing Incidence Vacuum Spectrograph," *J. Opt. Soc. Am.* **56**, 665 (1966).

GERHARZ, R., "The Crossed Grating Method for Order-Separating Spectra," *J. Quant. Spectr. Rad. Trans.* **6**, 59 (1966).

ZALUBAS, R., *New Description of Thorium Spectra*, Natl. Bur. Std. Monograph 17 (U.S. Government Printing Office, Washington, D. C., 1960).

GIACCHETTI, A., *Averages of Interferometric Measurements of Thorium Lines*, Argonne Natl. Lab. Report 7209, May 1966.

VALERO, F. P. J., "Thorium Lamps and Interferometrically Measured Thorium Wavelengths," *J. Opt. Soc. Am.* **58**, 484 (1968).

————, "Improved Values for Energy Levels, Ritz Standards, and Interferometrically Measured Wavelengths in Th I," *J. Opt. Soc. Am.* **58**, 1048 (1968).

GOORVITCH, D., F. P. J. VALERO, and A. L. CLÚA, "Interferometrically Measured Thorium Lines between 2747 and 4572 Å," *J. Opt. Soc. Am.* **59**, 971 (1969).

TOMKINS, F. S., and M. FRED, "Wavelength Measurements with a Concave Grating Spectrograph," *Appl. Opt.* **2**, 715 (1963).

STEINHAUS, D. W., R. ENGLEMAN, JR., and W. L. BRISCOE, "An Automatic Comparator for Measurement of Spectra," *Appl. Opt.* **4**, 799 (1965).

Index